T0092621

Advancing Edge Artificial Intelligence
System Contexts

RIVER PUBLISHERS SERIES IN COMMUNICATIONS AND NETWORKING

Series Editors:

ABBAS JAMALIPOUR
The University of Sydney, Australia

MARINA RUGGIERI
University of Rome Tor Vergata, Italy

The "River Publishers Series in Communications and Networking" is a series of comprehensive academic and professional books which focus on communication and network systems. Topics range from the theory and use of systems involving all terminals, computers, and information processors to wired and wireless networks and network layouts, protocols, architectures, and implementations. Also covered are developments stemming from new market demands in systems, products, and technologies such as personal communications services, multimedia systems, enterprise networks, and optical communications.

The series includes research monographs, edited volumes, handbooks and textbooks, providing professionals, researchers, educators, and advanced students in the field with an invaluable insight into the latest research and developments.

Topics included in this series include:

- Communication theory
- Multimedia systems
- Network architecture
- Optical communications
- Personal communication services
- Telecoms networks
- Wi-Fi network protocols

For a list of other books in this series, visit www.riverpublishers.com

Advancing Edge Artificial Intelligence
System Contexts

Editors

Ovidiu Vermesan

SINTEF, Norway

Dave Marples

Technolution B.V., The Netherlands

Routledge
Taylor & Francis Group

LONDON AND NEW YORK

Published 2024 by River Publishers
River Publishers
Alsbjergvej 10, 9260 Gistrup, Denmark
www.riverpublishers.com

Distributed exclusively by Routledge
4 Park Square, Milton Park, Abingdon, Oxon OX14 4RN
605 Third Avenue, New York, NY 10017, USA

Advancing Edge Artificial Intelligence / by Ovidiu Vermesan, Dave Marples.

ISBN: 978-87-7004-102-7 (hardback)
          978-10-4002-704-2 (online)
          978-10-0347-871-3 (master ebook)

DOI: 10.1201/9788770041027

Dedication

"The test of a first-rate intelligence is the ability to hold two opposed ideas in the mind at the same time and still retain the ability to function."

- F. Scott Fitzgerald

"The world needs dreamers, and the world needs doers. But above all, what the world needs most are dreamers that do."

- Sarah Ban Breathnach

"The real question is, when will we draft an artificial intelligence bill of rights? What will that consist of? And who will get to decide that?"

- Gray Scott

"The greatest masterpiece in literature is only a dictionary out of order."

- Jean Cocteau

"There's a way to do it better. Find it."

- Thomas Edison

Acknowledgement

The editors would like to thank all the contributors for their support in the planning and preparation of this book. The recommendations and opinions expressed in the book are those of the editors, authors, and contributors and do not necessarily represent those of any organizations, employers, or companies.

Ovidiu Vermesan
Dave Marples

Contents

Ovidiu Vermesan, Kai vorm Walde, Roy Bahr, Cordula Conrady,
Janis Judvaitis, Gatis Gaigals, Tore Karlsen, Marcello Coppola,
and Hans-Erik Sand

4　Inside the AI Accelerators: From High Performance to Energy Efficiency　　　**87**

Ana Pinzari, Adrien Prost-Boucle, Christelle Rabache, and Frédéric Pétrot

5　Designing Lightweight CNN for Images: Architectural Components and Techniques　　　**105**

Lilian Hollard, Lucas Mohimont, and Luiz Angelo Steffenel

**6   Natural Language Conditioned Planning of Complex Robotics
Tasks   131**
*Toms Eduards Zinars, Oskars Vismanis, Peteris Racinskis,
Janis Arents, and Modris Greitans*

**7   An Overview of the Automated Optical Inspection Edge AI
Inference System Solutions   153**
Claudio Cantone and Alberto Faro

Preface

Taking the Next Steps on the Journey to Intelligent Pervasive Networked Systems

Given AI's astounding progress over the past few years, it is time to posit the next question: *How do we integrate AI into IoT devices at the edge of the network?*

The pre-question should, of course, be *why?* That is easy to answer: AIs are becoming pervasive, but conveying the data they need over networks is time-consuming, expensive, and potentially risky. Migrating them to the network edge mitigates those issues.

Edge AI marks a shift from traditional cloud-centric AI models to decentralised computing power embedded directly into edge devices.

Modern high-performance, low-power silicon makes the proposition to move these AIs into the devices themselves viable, even though we don't really have too much clarity yet on how that will be done, how the devices will be managed, or what the consequences for our networked systems architectures will be. In this book, we start the process of addressing those uncertainties.

In the following chapters, we start documenting the journey to address these questions, starting with considering the underpinnings of our current network technology in Chapter 1 before regarding how we can manage the lifecycle of AIs in IoT devices in Chapter 2. Chapters 3, 4 and 5 investigate how we might teach these AIs before Chapter 6 introduces how we might communicate with them (we can consider screens and keyboards depreciated in this brave new world). Chapters 7 and 8 present two example environments where such AIs will find application, and Chapter 9 addresses how they can explain their actions.

We don't have the answers to the big questions yet. If we did, we'd be off-creating VC-backed startups somewhere rather than coordinating research programs, but we know that we're at the start of the next chapter of a fascinating journey.

This book provides valuable insight to researchers working with edge AI technologies, machine and deep learning engineers, IoT designers, and intelligent systems developers looking to deploy intelligent solutions at the edge.

List of Figures

List of Tables

List of Contributors

Adrien, Prost-Boucle, *Institute of Engineering Univ. Grenoble Alpes, France*

Alberto, Faro, *Deepsensing, DEEPS, Italy*

Ana, Pinzari, *Institute of Engineering Univ. Grenoble Alpes, France*

Andrija, Neskovic, *Universität zu Lübeck, Germany*

Angelo, Genovese, *Università degli Studi di Milano, Italy*

Celine, Thermann, *Universität zu Lübeck, Germany*

Christelle, Rabache, *Institute of Engineering Univ. Grenoble Alpes, France*

Claudio, Cantone, *High Technology Systems H.T.S. srl, Italy*

Claus, Lenz, *Cognition Factory GmbH, Germany*

Cordula, Conrady, *IMST GmbH, Germany*

Daniel, Hirsch, *NXP Semiconductors, Germany*

Dinu, Purice, *Cognition Factory GmbH, Germany*

Elhadj, Doguech, *Université Polytechnique Hauts-De-France, France*

Fabio, Scotti, *Università degli Studi di Milano, Italy*

Falk, Hoffmann, *NXP Semiconductors, Germany*

Francesco, Barchi, *Universita di Bologna, Italy*

Frédéric, Pétrot, *Institute of Engineering Univ. Grenoble Alpes, France*

Gatis, Gaigals, *Institute of Electronics and Computer Science, Latvia*

Hans-Erik, Sand, *NXTECH AS, Norway*

Ihsen, Alouani, *Université Polytechnique Hauts-De-France, France*

Iyad, Dayoub, *Université Polytechnique Hauts-De-France, France*

Janis, Arents, *Institute of Electronics and Computer Science, Latvia*

Janis, Judvaitis, *Institute of Electronics and Computer Science, Latvia*

Kai, vorm Walde, *IMST GmbH, Germany*

Lilian, Hollard, *Université de Reims Champagne-Ardenne, France*

Lucas, Mohimont, *Université de Reims Champagne-Ardenne, France*

Luiz, Angelo Steffenel, *Université de Reims Champagne-Ardenne, France*

Marcello, Coppola, *STMicroelectronics, France*

Mladen, Berekovic, *Universität zu Lübeck, Germany*

Modris, Greitans, *Institute of Electronics and Computer Science, Latvia*

Oskars, Vismanis, *Institute of Electronics and Computer Science, Latvia*

Ovidiu, Vermesan, *SINTEF AS, Norway*

Pasquale, Coscia, *Università degli Studi di Milano, Italy*

Peteris, Racinskis, *Institute of Electronics and Computer Science, Latvia*

Rainer, Buchty, *Universität zu Lübeck, Germany*

Roy, Bahr, *SINTEF AS, Norway*

Ruggero, Donida Labati, *Università degli Studi di Milano, Italy*

Saleh, Mulhem, *Universität zu Lübeck, Germany*

Taha, Yassine Abidi, *Université Polytechnique Hauts-De-France, France*

Thorsten, Röder, *Cognition Factory GmbH, Germany*

Toms, Eduards Zinars, *Institute of Electronics and Computer Science, Latvia*

Tore, Karlsen, *ProLux AS, Norway*

Vincenzo, Piuri, *Università degli Studi di Milano, Italy*

List of Abbreviations

AC	Approximate Computing
AES	Advanced encryption standard
AIA	Artificial intelligence act
AI	Artificial intelligence
AOI	Automated Optical Inspection
AODV	Ad hoc on-demand distance vector
ASIC	Application-specific integrated circuit
B.A.T.M.A.N.	Better approach to mobile ad-hoc networking (protocol)
BLE	Bluetooth low energy
BW	Bandwidth
CMM	Coordinate mount metrology
CNN	Convolutional neural network
CPU	Central processing unit
CS	AOI Solution using cloud server receiving images from cameras for both learning and testing
CSS	Chirp spread spectrum
DAG	Directed acyclic graph
DL	Deep learning
DLG	Deep leakage from gradients
DNN	Deep neural network
DnC	Divide and conquer
DSR	Dynamic source routing
ES	AOI Solution using testing boards at the edge and cloud server receiving images from cameras for learning
FAN	Field area network
FCT	Functional test
FEC	Forward error correction

FL	Federated learning
FPGA	Field programmable gate array
GAM	Generalised additive models
GDPR	General data protection regulation
GPU	Graphical processing unit
GS	AOI Solution using GPU workstation receiving images from cameras for both learning and testing
GSA	Global sensitivity analysis
IoT	Internet of Things
ICT	In circuit test
IR	Intermediate representation
IS	AOI Solution using inspecting machine close to the conveyor belt
JIT	Just in time compilation
Lime	Local interpretable model-agnostic explanation
LLN	Low-power and lossy network
LoRaWAN	Long-range wide area network
LUT	LookUp table
M2M	Machine-to-machine
MAC	Multiply accumulate; Medium/media access control layer
MANET	Mobile ad-hoc network
MAPLE	Model agnostic supervised local explanation
MEMS	Micro-electromechanical system
ML	Machine learning
MLE	Mesh link establishment
MLP	Multi-layer perceptron
MPR	Multi-point relay
MRI	Magnetic resonance imaging
NLP	Natural language processing
OGM	Originator message
OLSR	Optimised link state routing
PCBA	Printed circuit board assembly, sometimes printed circuit board assembler
PDP	Partial dependence plot
PTQ	Post-training quantization
QAT	Quantization aware training
OTA	Over-the-air

PHY	Physical layer
QoS	Quality of service
RNN	Recursive neural network
RPL	Routing protocol for low-power and lossy networks
SDLC	Software development lifecycle
SHAP	Shapley additive explanation
SIMD	Single instruction multiple data
SMT	Surface mount technology
SoC	System on a chip
SPI	Solder paste inspection
TPU	Tensor processing unit
TTL	Time to live
TVM	Tensor virtual machine
WLAN	Wireless local area network
WMN	Wireless mesh network
WSN	Wireless sensor network
XAI	Explainable AI
XLA	Accelerated linear algebra
YOLO	You only look once - Object detection model known for its speed and accuracy.

1

Edge AI LoRa Mesh Technologies

Ovidiu Vermesan[1], Kai vorm Walde[2], Roy Bahr[1], Cordula Conrady[2], Janis Judvaitis[3], Gatis Gaigals[3], Tore Karlsen[4], Marcello Coppola[5], and Hans-Erik Sand[6]

[1]SINTEF AS, Norway
[2]IMST GmbH, Germany
[3]Institute of Electronics and Computer Science, Latvia
[4]ProLux AS, Norway
[5]STMicroelectronics, France
[6]NXTECH AS, Norway

Abstract

Intelligent connectivity at the edge combines wireless communication, edge artificial intelligence (AI), edge computing and internet of things (IoT) technologies to perform machine learning (ML) and deep learning (DL) on connected edge devices. Low latency, ultra-low-energy intelligent IoT devices with on-board computing, and a distributed architecture and analytics are essential to drive intelligent connectivity.

Intelligent wireless mesh technologies exploit multiple interconnected devices, or nodes, to create a distributed network integrated with edge AI analytics using ML and DL algorithms. In an intelligent wireless mesh network (WMN), each node has embedded intelligence and can communicate directly with its neighbouring nodes and transfer data efficiently to other nodes. Compared with traditional point-to-point wireless networks, the intelligent wireless mesh approach offers several advantages, including increased coverage, redundancy, scalability and resilience.

The convergence of multiple technologies (connectivity, edge AI, IoT, distributed architectures and federated learning) delivers intelligent edge

DOI: 10.1201/9788770041027-1   1

mesh communication systems that perform efficient connectivity by optimising data rates, coverage, energy, and interference.

This article overviews the latest advancements in edge AI long-range mesh technologies and applications, highlights state-of-the-art mesh communication requirements and implementations and identifies future research challenges and directions.

Keywords: mesh communication technologies, edge artificial intelligence, LoRaWAN, LoRa mesh.

1.1 Introduction

Star, tree and mesh networks are examples of topologies used in communication networks. Each is suitable for different application scenarios. An illustration of the different network architectures is shown in Figure 1.1.

Star networks are simple to set up and manage because they have centralised control points. However, this makes them more susceptible to single-point failures. Mesh networks offer high redundancy and self-healing (e.g., recovery from a link failure), making them more reliable and fault tolerant at the cost of increased complexity.

Wireless mesh technologies play an essential role in creating robust and flexible wireless networks that address modern connectivity challenges.

In a star topology, all nodes are directly connected to a single central root node, often referred to as a hub. Direct peer-to-peer communication is not supported; all nodes must communicate through this central hub.

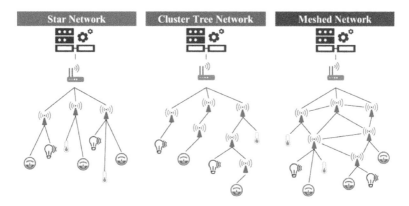

Figure 1.1   Network Topologies

Networks with cluster-tree topologies are divided into so-called clusters. Each cluster consists of a group of nodes connected to a local central node referred to as the cluster head. The cluster head coordinates the communication within its cluster. The tree terminology refers to the cluster heads which are organised in such a hierarchical structure. Communication from node to node may involve routing via multiple cluster heads.

Wireless mesh technologies use multiple interconnected nodes which can communicate directly with their neighbours nodes. This approach offers several advantages, including increased coverage, redundancy, resilience, and scalability.

Mesh communication technologies use distributed networking methods that typically create a decentralised and self-configuring network. Each node can also act as a repeater to extend network coverage and improve resilience.

The convergence of edge computing, edge AI, federated learning and IoT can create multi-dimensional architectures consisting of a wide range of heterogeneous entities with different sensing/actuating, connectivity, processing, and intelligence capabilities connected with applications in a dynamic mesh network linked by platforms and distributed services located at the edge level. Some of the technologies contributing to enhancing the capabilities of intelligent mesh connectivity include:

Edge AI - The deployment of AI algorithms and data processing capabilities directly on edge devices, rather than relying on centralised cloud servers, brings the following benefits:

- Real-time decision making – By processing data locally, AI models can make fast decisions without the latency of sending data to remote servers, enabling rapid responses to critical events.
- Data privacy and security – Edge AI reduces the need to transmit sensitive data to the cloud, increasing privacy and decreasing the consequence of data breaches.
- Bandwidth efficiency – Edge AI can filter and prioritise data before transmission, reducing bandwidth demands.

Federated Learning (also called collaborative learning) is a machine learning method in which edge devices collaboratively contribute to a global model while keeping their data locally. Federated learning can play a significant role in enhancing the combination of edge AI, IoT and communication technologies.

Privacy preservation: Federated learning avoids the transmission of raw data to a central server. This ensures that sensitive data remain on the edge devices, addressing privacy concerns and complying with data protection regulations.

Bandwidth efficiency: By training models locally on edge devices, federated learning reduces the need to send large amounts of data to the cloud for model training. This optimises bandwidth usage, making it more efficient for IoT devices with limited communication capabilities, such as long-range (LoRa)-based communication devices. LoRa is a wireless spread spectrum modulation technique derived from the chirp spread spectrum (CSS), which enables long-range communication between devices with low power consumption. The technology was initially developed by a company called Cycleo SAS and later acquired by Semtech Corporation [1], a semiconductor company specialising in analogue and mixed signal circuits.

Improved model performance: Federated learning allows IoT devices to continuously improve their local models. This can result in better model performance and adaptability over time, as each device benefits from the collective intelligence of the entire network.

Decentralised intelligence: Federated learning distributes intelligence across edge devices, promoting decentralised data processing and decision-making. This leads to increased resilience in the overall system.

Collaboration and knowledge sharing: By collaborating on model training, edge devices share knowledge and insight. This collaborative approach fosters rich and diverse learning experiences.

Reducing infrastructure costs: Federated learning reduces the need for large-scale cloud infrastructure for centralised model training. This results in cost savings in respect of data transmission and cloud computing resources.

Versatility and scalability: Federated learning can be adapted to many different edge devices and network architectures. It can scale efficiently making it suitable for IoT networks with diverse deployments and configurations.

Federated learning complements the combination of edge AI, IoT and LoRa by enhancing privacy, efficiency, model performance and collaboration. It empowers IoT networks with intelligent decision-making capabilities while respecting data privacy and promoting decentralised data processing.

Internet of Things (IoT) is related to the network of interconnected devices and sensors that collect, exchange, and analyse data. By integrating IoT with edge AI and LoRa technology, it becomes a powerful enabler across various domains:

- Remote monitoring and control – IoT sensors can collect data from different environments, enabling remote monitoring and control of processes, infrastructure, and assets.
- Predictive maintenance – IoT data, when combined with edge AI analytics, allows the prediction of equipment failures, optimisation of maintenance schedules and reductions in downtime.
- Energy management – IoT deployments combined with edge AI enable efficient energy management, waste reduction and improved urban services in smart city applications.

The combination of edge AI and mesh communication has several benefits, especially when infrastructure is impracticable or unavailable. Mesh networks enable flexible, reliable, and scalable networks. They are increasingly used in industrial IoT, energy, smart homes, agri-food and beverage, disaster recovery operations and smart city applications.

This chapter is organised into the following sections. Section 1 introduces the research area and the state of play of technology development. Sections 2 and 3 provide an overview of the state of the art of existing wireless mesh technologies and their primary functions, operating characteristics and actual advantages and disadvantages. Section 4 describes the LoRa wireless modulation technique and the long-range wide area network (LoRaWAN) technology, the main architectures, the architectural building blocks, and their characteristics. Section 5 covers enabling technologies (e.g., edge AI, edge computing, internet of intelligent things, artificial intelligence of things) and integration with LoRa mesh to enhance and optimise communication performance and mesh-based systems' collaborative and cooperative capabilities. Section 6 presents potential applications for LoRa mesh connectivity, edge AI and IoT systems and emphasises the requirements for intelligent communication and convergence with other technologies. Section 7 outlines the conceptual edge AI LoRa mesh device architecture. Section 8 analyses the state of play and future research directions and highlights several challenging open issues for intelligent edge LoRa meshes. Finally, Section 9 summarises the main points for discussion.

1.2 Overview of the State-of-the-Art Wireless Mesh Technologies

Meshes are networks that create a decentralised and robust structure where each node can communicate directly with neighbouring nodes.

Nodes are interconnected and, depending on the network topology, there can be multiple connection pathways for each node. Connections between nodes may be dynamically updated and optimised through a built-in mesh routing table. As nodes enter and exit the network, the mesh topology enables the nodes to reconfigure routing paths based on the new network configuration.

Mesh topology and ad-hoc routing assures stability in the face of changing communication conditions or node failure.

Mesh networks use a distributed approach, where each node can act as a repeater to extend network coverage and improve resilience. The critical characteristics of mesh communication technologies include:

- **Decentralisation** – mesh networks are not dependent on a single central point of control. Each node can communicate with its neighbour, allowing messages to bounce from one node to another until they reach their destination.
- **Self-configuration** – mesh networks are capable of self-organisation. When nodes are added or removed the network can dynamically reconfigure itself to accommodate these changes.
- **Redundancy and reliability** – due to their decentralised nature and self configuration capability, mesh network topologies are more resilient to node failure or network disruption.
- **Extended coverage** – mesh networks can cover an extended area by using multiple nodes as relays. This provides an advantage in cases when establishing a traditional infrastructure might be challenging or costly.
- **Ad-Hoc networking** – mesh communication technologies enable ad-hoc networking, where devices can spontaneously create a network without relying on pre-existing infrastructure.
- **Geographical scalability** – mesh networks can quickly expand their coverage by adding more nodes which do not need to be in direct communication.

1.2.1 Mesh components and roles

Wireless mesh networks usually consist of routers, nodes, and coordinators as described below:

- **Routers** – these devices form the backbone of a wireless mesh network. They are typically more powerful than simple nodes with enhanced processing capabilities and are responsible for routing data within the whole network. Mesh routers communicate with other routers and nodes in the network to forward data packets along the most efficient path to reach their intended destination.

- **Nodes** – these are individual devices connected to the mesh network. They can be computers, smartphones, sensors, IoT devices, or any other device capable of wireless communication. Mesh nodes are typically senders, receivers, or relay points. Unlike traditional networks, mesh nodes in a wireless mesh network can communicate directly with each other, creating multiple data transmission paths. This decentralised communication architecture enhances the network's reliability and overall performance.

- **Coordinators** – mesh coordinators are nodes with specialised roles in some wireless mesh network protocols. They act as central control points for the entire mesh network. A coordinator is responsible for managing and organising the network, assigning roles to other nodes (such as routers or end devices), and controlling aspects of the network's operation. They handle tasks like channel allocation, network formation, and security management. In some mesh network implementations, coordinators have a critical role in preserving the network's stability and performance. On one hand, central coordinators can offer efficient control and coordination; on the other hand, they can also become a single point of failure, potentially disrupting the entire network and compromising one of the key advantages of mesh topologies.

- **Decentralised functionality** – this approach eliminates the central mesh coordinator. Instead, the process of decision-making and control is distributed across multiple nodes. Nodes may possess a degree of autonomy, enabling them to make local decisions based on independent observations and interactions with neighbouring nodes. Local decisions collectively contribute to the overall behaviour of the network.

1.2.2 Wireless routing concepts

One of the key elements for wireless mesh communication, routing protocols are designed to enable communication and data exchange between devices in a wireless network. These protocols establish routes for data transmission and determine the best paths for information to flow from a source to a destination. The functions of a wireless routing protocol vary depending on the specific protocol used and the type of wireless network. We present a general overview of the common functionalities of these wireless routing protocols:

- **Neighbour discovery** – in wireless networks, devices must discover neighbours to establish direct communication links.
- **Route discovery** – when a device wishes to send data to another device, a route discovery process is initiated. During the process, the device searches direct links or for potential intermediate devices (routers) that can relay the data towards the destination. This process can involve broadcasting or multicasting route request packets to nearby devices to find potential routes.
- **Route maintenance** – once a route is established, the routing protocol is responsible for maintaining the health and stability of it. This includes monitoring the status of the intermediate devices along the path and detecting any changes, such as link failures or device mobility. If a route becomes unavailable, the routing protocol triggers a route repair process to find an alternative path.
- **Routing metrics** – wireless routing protocols use various metrics to determine the quality and efficiency of potential routes. Metrics include signal strength, link quality, distance, and available bandwidth. The routing protocol uses these metrics to select the preferred routes based on network conditions and requirements. The current battery state of a node may also be a metric to implement a kind of energy-balancing policy.
- **Data forwarding** – once a route is established, the data packets are forwarded from one router to the next until they reach their destination. Each router in the path makes a forwarding decision based on the routing table and the packet's destination address.
- **Adaptation to network changes** – wireless routing protocols are constructed to adapt to changes in the network topology, such as device mobility, link quality fluctuations, or node failures. They continuously monitor the network and adjust the routing paths to ensure reliable and efficient data transmission.

1.3 Routing protocols

Some standard wireless routing protocols, include Optimised Link State Routing (OLSR) [29][30][31][33], Ad hoc On-Demand Distance Vector

(AODV) [34][35], Dynamic Source Routing (DSR) [36][37] and Routing Protocol for Low-Power and Lossy Networks (RPL) [38][39][40]. Each protocol has specific features, advantages, with use cases tailored for different wireless networks and applications. There follow some details about the algorithms and their pros and cons.

1.3.1 Ad hoc on-demand distance vector (AODV)

AODV is a demand-driven reactive wireless routing protocol that establishes routes only when needed. When a source node requests to send data to a destination node, it initiates a route discovery process to find the most efficient path. The protocol uses sequence numbers to ensure loop-free routes and maintains a routing table to store information about discovered routes.

Pros:

- **Reduced overhead** – AODV minimises control message overhead by initiating route discovery only when necessary. This helps conserve network resources and reduces unnecessary traffic.
- **Loop-free routes** – using sequence numbers ensures that routes are loop-free, improving route stability and reliability.
- **Proactive link failure detection** – AODV employs proactive link failure detection to quickly identify failed links and initiate route repair, ensuring data continues to flow via alternative paths.
- **Scalability** – AODV performs well in moderately sized networks and maintains route information for frequently used paths, reducing route discovery latency.

Cons:

- **High latency for new routes** – AODV's on-demand route discovery process can introduce delays in finding a new route, especially in large networks or sparse topologies.
- **Route rediscovery** – several cases (link changes, node mobility, malicious nodes, battery depletion, network congestion or topology changes) lead to frequent route rediscovery, increasing control message overhead.
- **Suboptimal routes** – sometimes, AODV may not find the shortest path in specific network scenarios, leading to less efficient data transmission.

AODV balances control message overhead and route discovery latency, making it suitable for dynamic networks with changing topologies. However, its performance may vary depending on network size, mobility patterns, and the frequency of route changes.

1.3.2 Optimized link state routing (OLSR)

OLSR is a proactive routing protocol that uses a hybrid approach, combining both proactive and reactive mechanisms. It optimises link-state information exchange to minimise overhead while ensuring efficient route computation and maintenance. OLSR uses Multi-Point Relays (MPRs) to reduce control message flooding and speed up route discovery.

Pros:

- **Reduced control message overhead** – OLSR uses MPRs to limit the number of nodes participating in control message dissemination. This decreases control overhead and improves scalability, making it suitable for large networks.
- **Proactive and reactive hybrid approach** – OLSR combines proactive link-state information with reactive route discovery. It provides real-time responsiveness while minimising the amount of control traffic generated.
- **Loop-free routes** – OLSR guarantees loop-free routes and enhances route stability and reliability.
- **Fast route recovery** – MPRs and proactive topology updates enable quick route recovery and repair in case of link failures.
- **Better convergence** – OLSR converges quickly and efficiently, enabling devices to find optimised routes with lower latency.

Cons:

- **Memory and computation requirements** – OLSR requires storing and managing additional topology information due to MPRs. This imposes overhead which might be critical on devices with limited resources.
- **Increased initial setup overhead** – the initial setup phase in OLSR involves the exchange of control messages to determine MPRs which leads to higher overhead during network initialisation.
- **Relatively complex implementation** – compared to other protocols, the implementation and management of OLSR can be more complex due to its hybrid nature and the need to optimise MPR selection.

OLSR balances proactive and reactive mechanisms, making it suitable for dynamic networks with varying traffic patterns and topology changes. Its efficiency in controlling message overhead and quick route convergence makes it a viable choice for both small and large-scale wireless networks.

1.3.3 Dynamic source routing (DSR)

DSR is an on-demand routing protocol that establishes routes between nodes only when needed. When a source node requests to send data to a destination node, it initiates a route discovery process to find a path. The route discovery process is based on source routing, which includes the complete route in the data packet. Intermediate nodes use this route information to forward the packet to the next hop until it reaches the destination.

Pros:

- **Reduced overhead** – DSR minimises control message overhead since route discovery is initiated only when needed, conserving network resources and reducing unnecessary traffic.
- **Loop-free routes** – DSR ensures loop-free routes through sequence numbers and route caching, enhancing route stability and reliability.
- **Efficient source routing** – including the complete route in the data packet enables efficient source routing, eliminating the need for intermediate nodes to maintain routing tables.
- **Route repair** – DSR supports quick route repair in case of link failure, as the source node can initiate a new route discovery process to find an alternative path.

Cons:

- **Route discovery latency** – the route discovery process in DSR can introduce delays, especially in large networks or sparse topologies, as it requires time to find a route to a new destination.
- **Increased packet overhead** – including the complete route in the data packet leads to larger packet sizes, especially for long routes, resulting in increased packet overhead.
- **Route maintenance overhead** – frequent mobility or link changes can lead to higher route maintenance traffic, as DSR requires regular route updates to adapt to topology changes.
- **Source routing overhead** – While source routing eliminates the need for routing tables in intermediate nodes, it increases the size of data packets, which can be a concern for resource-constrained devices.

DSR offers a simple and efficient approach to routing in Mobile Ad-hoc Networks (MANETs), particularly for networks with moderate mobility and communication demands. Its reactive nature allows it to adapt to changing network conditions, while the use of source routing eliminates the need for routing tables in intermediate nodes. The trade-offs include potential overhead from route discovery and maintenance, which should be considered when selecting DSR as the routing protocol for specific MANET deployments.

1.3.4 Routing protocol for low-power and lossy networks (RPL)

RPL is a specialised routing protocol for low-power and lossy networks (LLNs) as commonly been in IoT and wireless sensor networks. RPL is a proactive routing protocol that forms a directed acyclic graph (DAG) to route data in LLNs efficiently. It organises devices into a tree-like structure, with a root node at the top. It optimises routes using objective functions based on specific metrics, such as energy efficiency or latency. RPL is tailored for devices with limited resources, making it well suited for battery-powered IoT devices that require reliable and energy-efficient communication.
Pros:

- **Energy efficiency** – RPL is designed to minimise energy consumption in resource-constrained devices. It optimises routes to ensure that energy is conserved during data transmission, thus prolonging the battery life of IoT devices and the entire IoT system.
- **Adaptability to LLNs** – RPL's tree-like DAG structure is well-suited for LLNs, where devices may have limited processing power and intermittent connectivity.
- **Objective function flexibility** – RPL allows network designers to choose different objective functions based on their specific requirements, such as energy efficiency, latency, or reliability.
- **Self-configuring and self-healing** – RPL networks can self-configure and adapt to changes in network topology, including the addition or removal of devices. It also supports self-healing, where the network finds alternative routes if link failures occur.

Cons:

- **Complex configuration** – configuring RPL for specific use cases can be complicated due to the various parameters and objective functions that must be considered. Proper tuning and optimisation may require expertise and considerable time.
- **Scalability for large networks** – while RPL performs well in small to medium-sized LLNs, it may face challenges in large networks, where the tree-like structure can lead to increased control traffic and reduced scalability.

- **Overhead in highly mobile networks** – in highly mobile LLNs frequent changes in the network topology may result in increased control message overhead as the network adapts to mobility.

Overall, RPL's focus on energy efficiency and adaptability to low-power and lossy networks makes it a strong choice for IoT and wireless sensor networks. It effectively addresses the unique challenges posed by resource-constrained devices, allowing them to form reliable and efficient communication links while optimising energy consumption. However, careful configuration and consideration of scalability in large networks are essential to ensure the protocol's effectiveness for specific deployment scenarios.

1.3.5 Wireless mesh protocols

Mesh communication technologies offer flexible, reliable, and scalable networking solutions, and several protocols include mesh topologies. A short overview of mesh protocols such as B.A.T.M.A.N., Bluetooth Mesh, OpenThread, Thread, ZigBee, Wi-Fi, Wi-SUN, WirelessHART, Z-WAVE and 6LoWPAN is presented before focusing on the LoRa mesh protocol and applications.

1.3.5.1 B.A.T.M.A.N

The protocol Better Approach To Mobile Ad-hoc Networking (B.A.T.M.A.N.) [11] is a multi-hop ad-hoc mesh network routing protocol where each node transmits broadcast or originator messages (OGMs) to notify neighbouring nodes about its presence. These neighbours re-broadcast the OGM.s based on specific rules to inform their neighbours about the presence of the original initiator. The network is steeped with OGM.s that are small, with a typical raw packet size of 52 bytes, including IP and UDP overhead. OGMs contain at least the originator's address, the address of the transmitting packet's node, a Time to Live (TTL) and a sequence number.

The approach of the B.A.T.M.A.N. algorithm is to divide the knowledge about the best end-to-end paths between nodes in the mesh to all participating nodes.

B.A.T.M.A.N. uses a proactive routing approach, which means it continuously maintains up-to-date routing information without waiting for a specific request to transmit data. Instead of relying on global routing tables, each node perceives and retains only the information about the best next hop towards all other nodes. Thereby the condition for overall network knowledge about local topology changes is unnecessary. Since wireless mesh networks are subject to frequent changes, B.A.T.M.A.N. is designed to be adaptive and capable of quickly reconfiguring routes when nodes join, leave, or move within the network.

The protocol also supports load balancing by distributing traffic across multiple paths to prevent congestion and optimise the overall network performance.

1.3.5.2 Bluetooth Low Energy

Bluetooth Low Energy (BLE) [13] is optimised for low power consumption to address small-scale consumer IoT applications. BLE is integrated into several IoT devices, and data is conveniently communicated to and visualised on smartphones. The Bluetooth Mesh specification aims to enable a scalable deployment of BLE devices.

BLE provides versatile indoor localisation features, and IoT beacon networks are used for different IoT service applications. BLE is incompatible/non-interoperable with Bluetooth, and a dual-mode device is required to achieve interoperability.

BLE uses multiple techniques to ensure low power consumption implementing the data protocol to create low-duty-cycle transmissions, combined with very low-power sleep modes.

Bluetooth Low Energy Mesh [12] protocol is a networking technology built on the BLE standard. It enables large-scale, reliable, secure communication between many devices, forming a mesh network. This mesh network allows devices to communicate with each other and extend the range of the network.

A device can have one or more logical elements in the Bluetooth Mesh network. Each element represents a specific functionality or component of the device, and each element is assigned a unique address within the network.

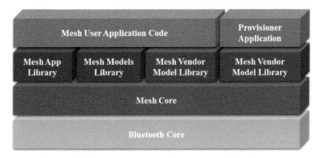

Figure 1.2   BLE Mesh Layered Architecture

Bluetooth Mesh devices use models to define their behaviour and capabilities. Models represent how a device handles messages, what types of messages it supports, and how it behaves in the mesh.

Provisioning is the process of securely adding a new device to the mesh network. Encryption keys and other necessary information are exchanged between the new device and the network during this process.

1.3.5.3 OpenThread and Thread

Thread [14] is a mesh networking low-power wireless protocol based on Internet Protocol version 6 (IPv6), designed to address the interoperability, security, power, and architecture challenges of the IoT. Thread utilises 6LoWPAN that employs the IEEE 802.15.4 wireless protocol with mesh communication. Thread is IP-addressable, with cloud access and advanced encryption standard (AES).

Thread uses a mesh network topology in the 2.4 GHz frequency spectrum, providing data rates of 250 kbps with a coverage range of 30 m. Security uses a 128-bit AES encryption system and the encryption cannot be disabled.

Thread utilises a network-wide key for inscription that is applied at the Media Access Layer (MAC). The key is employed as specified in IEEE 802.15.4. Attacks on Thread network originating over-the-air from outside the network are protected by IEEE 802.15.4 security mechanisms. The Thread network's nodes exchange frame counters with their neighbours via a Mesh Link Establishment (MLE) handshake. The protection against replay attacks is done via frame counters. Thread lets the application use various internet security protocols for end-to-end communication and can connect up to 250 devices.

OpenThread, released by Google, is an open-source implementation of Thread that implements all Thread networking layers (IPv6, 6LoW-PAN, IEEE 802.15.4 with MAC security, Mesh Link Establishment, Mesh Routing), device roles, and Border Router support.

1.3.5.4 ZigBee

ZigBee [15] is a short-range, low-power, wireless standard deployed in a mesh topology to extend coverage by relaying IoT sensor data over multiple sensor nodes.

The Zigbee standard works on the IEEE 802.15.4 physical radio specification and runs in unlicensed bands such as 2.4 GHz, 915 and 868 MHz.

Zigbee 3.0 sustains wireless networks' increasing scale and complexity and deals with extensive local networks of over 250 nodes. The data rates provided are 250 kbps (2.4 GHz), 40kbps (915 MHz) and 20kbps (868 MHz). Zigbee also handles the dynamic behaviour of the networks (with nodes disappearing, appearing, and re-appearing in the network topology) and permits orphaned nodes, resulting from the loss of a parent to rejoin the Zigbee network through another parent.

The self-healing structure of state-of-the-art Zigbee Mesh networks permits nodes to drop out of the network without disrupting internal routing. Zigbee supports over-the-air (OTA) upgrades during device operation and provides enhanced network security by employing a coordinator/trust centre, which creates the network and oversees the allocation of network and link security keys to joining nodes or distributed security where there is no coordinator/trust centre. The Zigbee router node can provide the network key to joining nodes.

1.3.5.5 Wi-Fi

Wi-Fi (IEEE/ISO/IEC 8802-11-2022) is a standard defining the characteristics of a wireless local area network (WLAN). The name Wi-Fi (short for "Wireless Fidelity") relates to the name provided by the Wi-Fi Alliance, formerly WECA (Wireless Ethernet Compatibility Alliance). This group assures compatibility between hardware devices that use the 802.11 standards. Wi-Fi networks must comply with the 802.11a-x specifications.

Wi-Fi mesh [16] protocol IEEE 802.11s creates a mesh network that extends Wi-Fi coverage over a larger area and enhances overall network performance and reliability. Traditional Wi-Fi networks are based on a single wireless access point (router) communicating directly with Wi-Fi-enabled devices. They may suffer from limited range and dead zones in larger spaces.

A Wi-Fi mesh network consists of multiple interconnected access points that work together to create a seamless and continuous network. These access points, often referred to as "nodes" or "mesh nodes", communicate with each other wirelessly, forming a self-healing network that can automatically reroute data packets to find the most efficient path to reach the destination device.

The system architecture for WLAN mesh network technology is described in IEEE 802.11 functional requirements and scope [17] and illustrated in Figure 1.3.

The functional blocks of the architecture include the following:

- The Mesh Topology Learning, Routing, and Forwarding block includes a function for discovering neighbouring nodes, a function for obtaining radio metrics, which deliver information on the quality of wireless links,

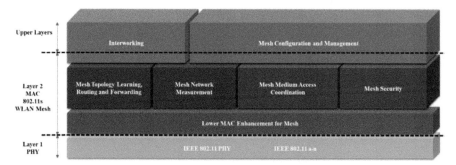

Figure 1.3 Wi-Fi Mesh Layered Architecture

a routing protocol for determining routes to transfer packets to their destinations using MAC addresses as identifiers, and a packet forwarding function. The routing protocol must use radio metrics and multiple frequency channels according to radio conditions to efficiently use radio resources.

• The Mesh Network Measurement block includes functions for calculating radio metrics used by the routing protocol and measuring radio conditions within the WLAN mesh network for frequency channel selection.

• The Mesh Medium Access Coordination block contains functions for preventing degraded performance due to hidden and exposed terminals, procedures for performing priority control, congestion control, and admission control, and a function for achieving spatial frequency reuse.

• The Mesh Security block comprises security functions (e.g., WLAN security schemes defined by the IEEE 802.11 standard) for protecting data frames carried on the WLAN mesh network and management frames used by control functions such as routing protocol.

• The Interworking block implements the function that supports WLAN mesh network to conform to IEEE 802 network architecture and connect to other networks by implementing a transparent bridge function enforced in the mesh portal situated at the network boundary. Each WLAN mesh network must operate as a broadcast network to deliver forwarded packets to all terminals connected to the LANs.

• The Mesh Configuration and Management block comprises a WLAN interface for the automatic setting of each mesh point's RF parameters (transmit power, frequency channel selection, etc.) and quality of service (QoS) policy management.

Wi-Fi mesh protocol is designed to address the limitations of traditional Wi-Fi networks, making them ideal for large homes, offices, or public spaces where extended coverage and high-performance connectivity are required.

1.3.5.6 Wi-SUN

Wi-SUN [18] stands for Wireless Smart Ubiquitous Network and is a mesh network protocol developed by Wi-SUN Alliance. Wi-SUN is one of the most popular IPv6 sub-GHz mesh technologies for smart utility and smart city applications. The target networks are named Field Area Networks (FANs), and they deliver a communications infrastructure for large-scale outdoor networks, usually outdoor IoT devices. FANs let industrial devices such as smart meters and streetlights interconnect onto one common network.

Wi-SUN is based on the IEEE 802.15.4g standard for the physical layer (PHY) and the IEEE 802.15.4e standard for the medium access control layer (MAC). It supports multiple data rates and frequency bands to meet regulatory requirements worldwide.

Wi-SUN makes interoperable, multi-service, secure wireless mesh networks available to service providers, utilities, municipalities/local governments, and other businesses. Wi-SUN can be used in various line-powered and battery-powered applications for large-scale outdoor IoT wireless communication networks. With the help of Wi-SUN, developers can add new features to existing infrastructure platforms by extending open standard internet protocols (IP) and APIs. With its long-range capabilities, high data throughput, and support for IPv6, Wi-SUN is designed to scale and makes wireless infrastructure easier for commercial applications and the development of smart cities.

1.3.5.7 WirelessHART

WirelessHART [19][20] is a process automation application wireless communications protocol that provides wireless capabilities to extend Highway Addressable Remote Transducer (HART) by keeping compatibility with existing HART commands, tools, and devices.

The architecture of the WirelessHART protocol stack according to the OSI 7-layer communication model is illustrated in Figure 1.4.

The WirelessHART protocol stack addresses five layers: physical layer, data link/MAC layer, network layer, transport layer and application layer. A central network manager is added for arbitrating the communication schedule and manage the routing.

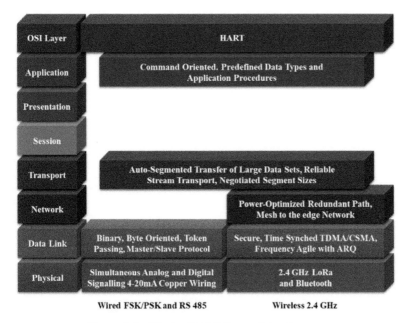

Figure 1.4 WirelessHART Protocol Architecture

WirelessHART uses mesh networking technology by design, where each device in a mesh network can act as a router for messages from other devices. This widens the range of the network and gives redundant communication routes to extend reliability in challenging radio environments encountered in process facilities [21][22][23]. Networks can scale up to 1000 nodes, but latency can be long and nondeterministic because transmissions occur only within an allocated time slot, and retransmissions are minimised.

Each WirelessHART network contains three major components:

- Wireless field devices that are connected to process or manufacturing equipment.
- Gateways that communicate among devices and on-premises host applications connected to high-speed backbone or other communications networks.
- A Network Manager configures the network, schedules communications between devices, monitors network health, and manages message routes. The Network Manager can be embedded into gateways, host applications, or process automation controllers.

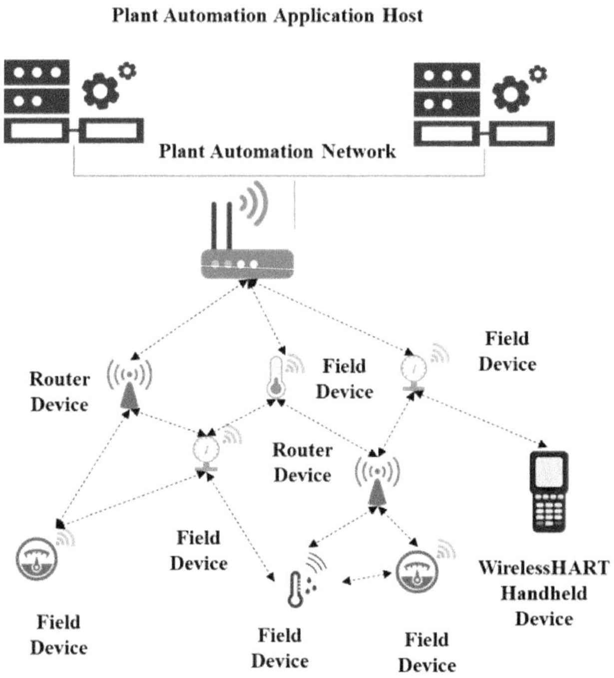

Plant Automation Application Host

Figure 1.5   WirelessHART Mesh Networking

WirelessHART supports different messaging modes, such as one-way publishing of process and control values, spontaneous exception notification, ad-hoc request/response, and auto-segmented block transfers of large data sets, to provide flexibility to meet different application requirements. These features enable communications to be tailored to the application's needs, lowering power consumption and overhead.

The WirelessHART mesh networking topology applied to an industrial plant use case is illustrated in Figure 1.5.

WirelessHART is used in industrial environments that require security to provide the highest levels of protection to the network and data. Security includes encryption and authentication.

1.3.5.8 Z-WAVE

Z-Wave [24][25] is the wireless technology for secure, trusted home applications like home appliances, lighting control, security systems, garage door openers, thermostats, windows, locks, etc.

It is a mesh network low-energy wireless communications protocol used in systems controlled via the Internet and locally through devices or a Z-Wave gateway or central control device serving as hub controller and portal.

The Z-Wave Alliance [26] demands the mandatory implementation of the Security 2 (S2) framework on all devices receiving certification. Z-wave delivers packet encryption, integrity protection and device authentication services. End-to-end security is provided at the application level (communication using command classes). It has an in-band network key exchange and AES symmetric block cipher algorithm using a 128-bit key length.

Products using Z-Wave mesh protocol are interoperable and communicate with each other regardless of brand or platform, and the Z-Wave mesh networks become more reliable as more devices are added (e.g., a Z-Wave network with 100 devices is more reliable than a Z-Wave network with 30 devices). Z-Wave's interoperability at the application layer assures that Z-Wave devices share information and allows all Z-Wave hardware and software to work together.

Z-Wave uses the unlicensed industrial, scientific, and medical (ISM) band and operates at 868.42 MHz in Europe and 908.42 MHz in the US. Z-Wave delivers data rates of 9.6 kbps and 40 kbps, with output power at one mW.

Z-Wave range between two nodes is 100 m in an outdoor, unobstructed setting. For in-home applications, the range is 30 m for no obstructions and 15 m with walls in between.

1.3.5.9 6LoWPAN

6LoWPAN [27][28] itself is not a mesh protocol; it is an open standard defined in RFC 6282 by the Internet Engineering Task Force (IETF) for a network where every wireless network node is battery-powered and has a IPv6 address. Thus, a set of local nodes can make a wireless mesh network.

6LoWPAN defines how to run IP version 6 (IPv6) over low data rate, low power, and small footprint radio networks (LoWPAN) as typified by the IEEE 802.15.4 radio [28].

IP addresses may be static or dynamic if a network node that can issue IPv6 addresses is acting as or like a Dynamic Host Configuration Protocol (DHCP) server. For IoT networks, it is typical to have a node connected to both WLAN and LAN that performs the gateway functions to collect local data and control local nodes. If local 6LoWPAN demands such a functionality, it typically performs the DHCP server functions too.

1.4 LoRa and LoRaWAN Technology

LoRa and LoRaWAN are related but distinct technologies used together to create long-range, low-power wireless communication networks for the IoT and other edge applications.

1.4.1 LoRa physical layer

LoRa operates in the sub-GHz ISM bands, such as 433MHz, 868 MHz (Europe) or 915 MHz (North America).

Semtech has released a LoRa chipset operating at the 2.4 GHz frequency band, which is globally available with km-range capabilities, enabling region-independent hardware design chipsets [3][4].

LoRa, compared with other technologies operating in the 2.4 GHz band, such as Wi-Fi and Bluetooth, offers several significant advantages in range and power consumption in comparison with other existing techniques.

The BLE standard range is from 50 m indoors to 165 m outdoors, and the maximum range of 2.4 GHz Wi-Fi networks typically reaches around 100 m. LoRa's outdoor range is more than five times the outdoor range of BLE, and more than eight times typical IEEE 802.11 networks.

LoRa modulation is able to offer a higher receiver sensitivity and robustness against noise and interference. Some of the specific details will be explained in the next sub-chapters.

Chirp Spread Spectrum Modulation (CSS)

LoRa modulation uses a form of chirp spread spectrum modulation, where the transmit signal frequency varies continuously over time. Instead of transmitting data on a fixed carrier frequency, LoRa uses chirp signals that start at one frequency and sweep across the spectrum. The LoRa chirping signal sequence makes LoRa signals robust against narrowband interference because the signal energy is spread over a wider frequency range.

Symbols and Data Rate

LoRa allows to adapt the number of bits per symbol according to the signal-to-noise ratio available over the link. Long range is achieved by reducing the number of bits per symbol, increasing the amount of energy per bit, and thus reducing the resulting bit rate.

Spreading Factor (SF)

The spreading factor (SF) is a critical parameter in LoRa modulation that determines the signal's robustness and range. The SF defines the rate at which the chirp signal spreads across the frequency spectrum and the amount of (potential) processing gain on receiver side.

Higher SF results in a lower data rate but better resistance to interference and an extended communication range. Conversely, a lower SF provides a higher data rate but with reduced range and increased susceptibility to noise.

Signal Bandwidth (BW)

The bandwidth of the LoRa signal also influences communication performance. LoRa modulation can operate in different bandwidths, typically 125 kHz, 250 kHz, or 500 kHz for sub-GHz LoRa.

A wider bandwidth allows for higher data rates but may reduce the communication range. Narrower bandwidths, on the other hand, result in lower data rates but offer increased range and better interference immunity.

Reception and Demodulation

On the receiver side, LoRa demodulation involves analysing the received chirp signal to decode the transmitted symbols. The receiver can determine the transmitted symbols and extract the original data by comparing the received signal with predefined chirp sequences.

Forward Error Correction (FEC)

In addition to the modulation scheme, a forward correction algorithm with several code rates can be applied, which enables the receiver to recover corrupted bits. This feature helps to decrease the number of packet retransmissions in noisy environments.

Sub-GHz Frequency Bands

The license-free sub-GHz ISM band allows transmitting within fixed defined frequency bands which vary depending on the region.

In this context, it is not possible to use the same type of radio hardware equipment because the used frequencies significantly impact the used chips, antenna matching circuits and the connected antennas.

The combination of a robust wireless transmission scheme with long-range capabilities and a low power footprint makes the LoRa technology ideal for battery powered IoT devices that can last up to 10 years.

The LoRa technology became public combined with the first LoRa radio modules and the so-called LoRaMAC protocol, today known as LoRaWAN protocol and defined within the LoRaWAN standard.

The following subchapters outline the most compelling aspects of the standard.

Table 1.1   Frequency Band Overview

No.	Region	Frequency Band
#1	Europe	863 MHz – 870 MHz
#2	Europe	433,05 MHz - 434,79 MHz
#3	North America	902 MHz– 928 MHz
#4	China	470 MHz – 510 MHz
#5	Korea	920 MHz – 925 MHz
#6	Japan	920 MHz – 925 MHz
#7	India	865 MHz – 867 MHz

1.4.2 LoRaWAN protocol

The LoRaWAN protocol defines methods, packet formats and LoRa physical layer radio parameters to ensure interoperability between IoT end devices and a given network infrastructure. The LoRaWAN standard itself is maintained by the non-profit association the LoRa Alliance [2].

The standard defines a system architecture consisting of at least three different component types with different roles and responsibilities.

The composition of end devices, gateways, and a central network server enables applications to create a star-of-star network topology.

LoRaWAN End Devices

These are typically sensors or actuators that need to communicate wirelessly over large distances through the LoRaWAN Link Layer protocol, formerly known as LoRaMAC protocol.

LoRaWAN Gateways

Gateways operate as intermediate devices with less intelligence. They relay the uplink and downlink messages between end devices and the network server using different TCP/IP-based protocols. A network can consist of several gateways.

LoRaWAN Network Server

The network server includes all the intelligence for controlling the radio network resources, e.g., network access, a security parameter, spreading factors (adaptive radio data rates) etc.

The network server is connected to all gateways and the application server, which hosts the application data and business logic. Suitable TCP/IP-based protocols typically handle these connections.

LoRaWAN allows IoT devices to transmit data over long distances to LoRaWAN gateways, which act as intermediaries between the end devices and the network server. LoRaWAN's key features are:

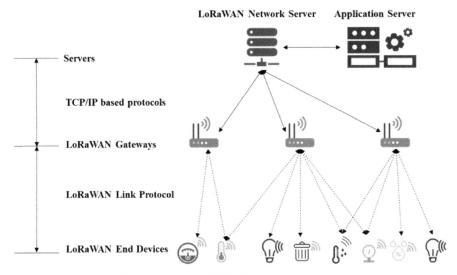

Figure 1.6 LoRaWAN Network Architecture

- **Low power** – LoRaWAN is designed to operate with low-power IoT devices, enabling long battery life for sensors and devices.
- **Wide area coverage** – LoRaWAN provides wide area coverage by leveraging the long-range capabilities of LoRa.
- **Public or private networks** – LoRaWAN can be deployed in public networks managed by network operators or in private networks operated by organisations.
- **Security** – LoRaWAN incorporates several security mechanisms, including end-to-end encryption and device authentication, to ensure secure data transmission.
- **Adaptive data rate** – LoRaWAN supports adaptive data rates, allowing devices to adjust their transmission speed based on the quality of the communication link, ensuring efficient data transfer.

LoRa and LoRaWAN form a powerful combination for creating efficient and scalable IoT communication networks. LoRaWAN defines a communication protocol and network architecture for IoT low-power wide area networks (LPWANs) and is designed to address the requirements for low power consumption (i.e., long battery life), long-range, and variable data rates (0.3 kbps – 50 kbps) while maintaining low operating and deployment costs.

1.4.3 2.4 GHz LoRa

In addition to sub-GHz LoRa, Semtech has developed a transceiver circuit with LoRa modulation for the 2.4 GHz ISM band. Compared to the sub-GHz solution this radio enables additional applications with diverse requirements.

The 2.4 GHz LoRa might be more suitable for applications operating in urban environments with higher device density, but covering shorter distances. On the other hand, sub-GHz LoRa is well-suited for applications needing extended range and better penetration of obstacles. Table 1.2 offers a brief comparison of the two radio technologies.

The integration of 2.4 GHz LoRa and a mesh protocol stack holds the potential to enhance the capabilities of edge AI-enabled IoT applications, particularly in terms of range coverage, network density, and robustness against single points of failure.

Table 1.2   Frequency Band Overview

Aspect	Sub-GHz LoRa	2.4 GHz LoRa
Frequency Band	433 MHz, 868 MHz, 915 MHz, depending on region / country	2.4 GHz Worldwide available
Range	Longer range	Shorter range
Penetration	Better penetration of obstacles	Lower penetration
Susceptibility to Interference	Lower	Higher due to higher signal channel bandwidth and multiple usage of the 2,4 GHz ISM band
Applications	Agriculture, rural areas, wide-area IoT networks	Smart Cities, densely populated areas, short-distance IoT networks
Interference Potential	Lower potential	Higher potential
Network Density	Lower density networks	Higher density networks
Tx Limits	Duty Cycle Limit 0.1%, 1%, 10% depending on sub-band	Unlimited
Bandwidth	125 kHz, 250 kHz, 500 kHz	203 kHz, 406 kHz, 812 kHz, 1625 kHz
Data rate	0.3 kbps – 0.9 kbps	0.2 kbps - 203 kbps

1.5 LoRa Mesh and Enabling AI Technologies

The convergence of technologies (including edge AI, IoT, distributed architectures, and federated learning) results in intelligent edge mesh communication systems performing efficient connectivity by optimising data rates, coverage, energy, and interference. LoRa when combined with edge AI and IoT, enhances connectivity and enables novel use cases:

- **Comprehensive area coverage** – LoRa's long-range capabilities allow devices to communicate over several kilometres, making it suitable for large-scale IoT deployments in smart agriculture, asset tracking, and environmental monitoring.
- **Energy efficiency** – LoRa devices consume very little power, making them ideal for battery-operated IoT sensors and devices, which can operate for extended periods without frequent battery replacements.
- **Low cost and scalability** – LoRa's low infrastructure cost and simple deployment enable cost-effective and scalable IoT solutions across diverse environments.

The Figure 1.7 illustrates a typical mesh topology with end nodes and gateways offering AI. For tasks like secure device enrolment, automatic firmware deployments or additional system monitoring a single or multiple application servers can be connected by wired or wireless IP based communication links. By combining edge AI, IoT, and LoRa, adopters can benefit from improved data rates, reduced latency, increased efficiency,

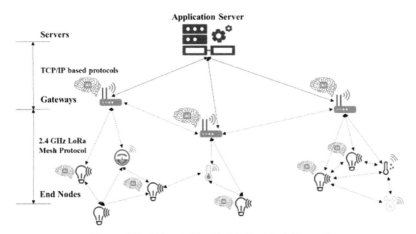

Figure 1.7 Edge AI Enabled LoRa Mesh Network

and cost-effectiveness. This convergence opens opportunities for innovation, automation, and optimisation across various sectors.

1.6 Applications for LoRa Mesh

LoRa mesh networks offer a versatile and reliable solution for applications that require low-power and extended-range wireless communication. LoRa mesh networks are suited for the following applications:

Industrial Automation: In industrial settings, LoRa mesh networks can be deployed for machine-to-machine (M2M) communication, asset tracking, and control systems. They enable monitoring and control of equipment and processes with extended-range.

Building Management Systems: LoRa mesh networks can be employed to optimise energy consumption in commercial buildings by managing lighting and other energy-related equipment more efficiently. However, it can be argued to what extent it remains energy efficient to reach indoor end nodes from an outdoor base station.

Smart Metering: LoRa-based intelligent metering systems can enable utilities to remotely monitor and manage energy, water, and gas consumption in residential and industrial settings.

Wireless Sensor Networks (WSNs): LoRa is a popular choice for creating WSNs, where many battery-powered sensors communicate with a gateway for data collection and analysis.

Smart Agriculture: LoRa mesh networks can be deployed in agricultural settings to monitor soil conditions, automate irrigation systems, and track livestock.

Lighting Control: LoRa can be used in wireless lighting control systems, enabling users to create adaptive and energy-efficient lighting environments.

Environmental Monitoring: LoRa mesh networks can be employed for monitoring environmental parameters, such as air quality, temperature, and humidity, in smart cities or remote areas. Furthermore, those networks can aid in predicting critical situations such as fires, floods, or earthquakes.

1.7 Conceptual Edge AI and LoRa Mesh Device Architecture

This chapter outlines a possible device architecture which integrates AI and 2.4 GHz LoRa Mesh technologies.

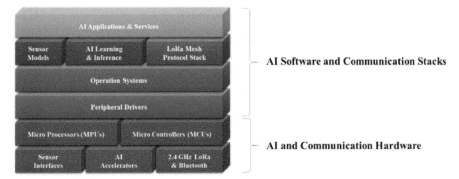

Figure 1.8   Conceptual Edge AI and LoRa Mesh Device Architecture

The purpose of this architecture is to provide a foundational framework for designing and implementing edge-devices with respect to the hardware and embedded software aspects.

The subsequent subsections should provide more detailed explanations of the provided building blocks, starting from the bottom, which includes the hardware-related units.

1.7.1 Sensor and interfaces

Typical IoT end devices include sensors (or actuators) that are connected via serial interfaces such as UART, SPI, or I2C to embedded microcontrollers running corresponding sensor drivers. More sophisticated devices may feature camera interfaces for image processing or display interfaces to connect displays that provide complex visual feedback to users. Consequently, the selection of microcontrollers/processors, sensor interfaces, power supplies, and connectors greatly depends on the specific application requirements. Designers and engineers must consider these factors when developing either a dedicated device or a multipurpose edge AI computing platform.

1.7.2 AI accelerators

Compared to pure software solutions, AI hardware accelerators offer better computational performance with a lower energy consumption footprint due to their parallel architecture. AI accelerators are designed for deep learning (DL) neural network computations and machine learning (ML) applications.

1.7.3 2.4 GHz LoRa and Bluetooth radios

The integration of 2.4 GHz LoRa and Bluetooth radio technologies can be achieved using modules that include their own microcontroller running the corresponding protocol stack. Such modules typically offer serial interfaces like UART or SPI for configuration, control, and data transfers. While 2.4 GHz LoRa is primarily used for long-range data exchanges within the application, short-range Bluetooth can be used for tasks such as single device maintenance and firmware updates. This can be accomplished through smartphones, tablets, or other portable computers that have Bluetooth available as a standard connectivity service.

1.7.4 Microcontrollers and microprocessors

These units are available from various manufacturers, offering a wide range of processing capabilities, including single-core and multi-core devices, as well as various memory and interface options. Microprocessor systems are typically capable of running embedded Linux, providing enhanced flexibility in choosing an appropriate programming language with higher abstraction and extensive library support. Microcontrollers are more likely to run smaller operating systems like FreeRTOS or proprietary ones, often with varying levels of real-time support and are directly connected to sensors and actuators.

Arm-based architectures with AI/ML-optimised cores support the development of lightweight microcontrollers with embedded coprocessing to optimise overall processing capability, local analytics, and power consumption. The edge AI methods, techniques, frameworks, and tools enable the embedded design to develop, train, optimise and deploy edge AI models on microcontroller-based hardware.

1.7.5 Peripheral driver

The connection between hardware and software is typically established through peripheral drivers. These drivers offer an interface for the higher layers of embedded software and ensure secure control and configuration of the underlying hardware units. In the case of operating systems like Linux, such drivers must adhere to specified interfaces and be implemented according to predefined rules. Additionally, in smaller microcontroller-based systems, similar driver software has been developed for the same purpose.

1.7.6 Operating systems

An operating system acts as an intermediary between hardware and embedded software applications. It manages and coordinates various hardware and software components to provide a stable and efficient environment for middleware and application software to run on a device. The choice of the operating system is, like hardware selection, significantly dependent on application requirements. Furthermore, it must be compatible with the selected hardware to support the lower-level peripheral drivers and interfaces.

1.7.7 Sensor models

A sensor model is a representation of how a sensor behaves and interacts with the environment it is monitoring. The model is a mathematical or computational description that helps understand and predict the relationship between the input (physical quantity being sensed) and the output (measurement or signal generated by the sensor).

Sensor models are used for various purposes, including:

- Simulation – they can be used to create virtual sensor behaviours in software simulations, allowing engineers to test systems before physical implementation.
- Calibration – sensor models help in calibrating real sensors by understanding how their measurements correspond to actual physical values.
- Data Fusion – when multiple sensors are used to gather information, their models can help combine and interpret the data accurately.
- System Design – in designing complex systems, sensor models aid in selecting appropriate sensors and understanding their integration.
- Fault Detection – deviations between actual sensor outputs and model predictions can indicate sensor malfunctions.

Sensor models can be as simple as linear equations or as complex as sophisticated computational simulations. They consider various factors that affect sensor behaviour, such as noise, sensitivity, non-linearity, temperature dependence, and more. By having an accurate model, engineers can improve the reliability and accuracy of systems relying on sensor data.

1.7.8 AI learning and inference

This building block includes the two fundamental aspects of an AI enabled edge device.

- AI Learning – this is the process in which AI systems gain knowledge and insights from data. It employs algorithms to identify patterns and learn how inputs relate to desired outputs. There are different types of AI learning, including supervised, unsupervised, and reinforcement approaches.
- AI Inference – this is the phase when a trained AI model is used to make predictions or decisions based on available data. This is the practical application of what the AI system has learned before.

1.7.9 2.4 GHz LoRa Mesh Protocol Stack

The LoRa Mesh Protocol Stack encompasses the functionalities to enable end-to-end communication within a wireless mesh topology. This includes tasks such as neighbour and route discovery, packet forwarding, route adaptation and maintenance, device management, and medium access control. Additionally, specific metrics and interfaces may be exposed to the embedded AI unit, enhancing adaptive routing algorithms through AI-based methods and techniques.

1.7.10 AI applications and services

This upper layer encompasses specific aspects and services tailored to a particular distributed edge AI-enabled application. The associated software components within this layer utilize middleware layer components at the highest available abstraction level to meet the application's specific functional and non-functional requirements.

1.8 Challenges and Future Research Directions

Built-in edge AI and wireless mesh connectivity capability that integrates processing units with AI-based capabilities, multiprotocol communication wireless modules for real-time monitoring and high-performance micro-electromechanical systems (MEMS) accelerometer sensors extend the functionalities and features of intelligent edge devices. This facilitates data aggregation, integration, and processing.

Building AI into wireless edge devices and sensors allows edge devices to learn and infer. Inference and decision making are performed within the edge device based on data collected through its sensors.

Long-range mesh network designs with edge AI capabilities enable effective monitoring through infrequent data updates communicated over long distances.

The LoRa mesh network can include security mechanisms while maintaining a low-energy profile for battery-powered edge sensors.

Lightweight authentication and encryption techniques can avert spoofing and provide confidentiality in message exchanges between edge nodes and the base station.

Updates can be performed using GPS-enabled time synchronisation and a concurrent transmission property inherent to LoRa.

An overview of the primary challenges and future research directions for edge AI and wireless LoRa mesh connectivity is presented in the next paragraphs.

Various wireless routing protocols, such as AODV, OLSR, DSR and RPL, face different challenges depending on the specific characteristics of the networks in which they are deployed. The following are some common challenges that these protocols frequently encounter:

- **Scalability** – all of these protocols need to scale with the increasing number of nodes in a network. As the network grows, more routing information must be managed and distributed. This can lead to increased overhead and longer route discovery times, especially for protocols based on proactive topology updates.
- **Mobility** – in wireless networks, devices can move frequently or follow unpredictable patterns. Protocols must be able to adapt to these changes and maintain efficient routes, even for mobile devices.
- **Connectivity** – fluctuations and interferences – Wireless networks are susceptible to connectivity fluctuations, interferences, and signal attenuations. Routing protocols must cope with these variations to provide stable and reliable routes.
- **Energy efficiency** – energy efficiency is crucial in IoT networks and battery-operated devices. Routing protocols should be designed to minimise energy consumption and maximise battery life.
- **Security** – wireless networks are vulnerable to security threats, such as man-in-the-middle attacks and routing manipulation. Routing protocols must rely on other mechanisms to secure communication and ensure the integrity of routing information.

- **Overhead and latency** – routing protocols generate additional overhead in the network to distribute and update routing information. This overhead can reduce the available bandwidth and lead to higher latency and increased energy consumption.
- **Complexity** – some routing protocols can be complex, especially when optimised for specific use cases. The implementation and management of such protocols can be challenging.
- **Interoperability** – in some cases, wireless networks must communicate with different devices and technologies from other vendors or protocols. Ensuring interoperability between different protocols can be a challenge.

It can be a challenge to find the appropriate edge AI learning techniques and AI input parameters when combining communication protocols with AI-based methods at the application level to enhance the overall performance of the wireless network itself.

These challenges are crucial when selecting and implementing a routing protocol for a wireless network. The routing protocol must meet the specific requirements of the network and the characteristics of the connected devices to ensure optimal performance and reliability.

The challenges for federated learning systems are potentially related to wireless communication efficiency, platform and sub-system heterogeneity, data heterogeneity, and protection of privacy [41][42][43]:

- **Wireless communication efficiency** – federated networks can include many edge nodes/devices, and the communication latency in the network may be larger than the time for computations carried out locally at the edge nodes/devices. As for all wireless networks, bandwidth is limited depending on the wireless technology solution used. Efficient communication strategies are needed to reduce the size and number of messages transmitted, such as the communication rounds constituting the training cycles, which are typically repeated iteratively until the global model converges and the targeted accuracy is achieved. To increase communication efficiency, local updates carried out in parallel on the nodes/devices for each communication round can reduce the total number of communication rounds. The size of messages transmitted can be reduced by using model compression methods, such as subsampling and quantification, and latency and bandwidth challenges can be reduced by decentralised topologies and training.

- **Platform and sub-system heterogeneity** – the federated network is a heterogeneous system typically without inherent seamless properties. The system may be challenged by different communication protocols, variations in hardware capabilities (e.g., processor units and memories) and various restrictions on energy consumption. When many edge nodes/devices are included in a system or its sub-systems, node/device fault tolerance properties are essential for the case of node loss (e.g., communication failure or power limitations) during a training/learning iteration. To reduce the possible adverse effects of heterogeneity, parallel iterative operations can be facilitated by asynchronous communication, and the number of nodes/devices participating in each communication round for training/learning can be increased and/or selected by active node/device sampling to maximise the aggregated node/device update within a defined timeframe; the effect of dropout can also be reduced/eliminated by implementing fault tolerance solutions that facilitate redundancy.

- **Data heterogeneity** – the data collected/generated from a potentially large number of edge nodes/devices in a federated network may be heterogeneous because of differences in populations, samples, and results. That is; the data used for modelling and analysing are usually not uniformly distributed across the edge nodes/devices, and variations in data types, attributes, data labelling, data points and data refresh rates, challenge the training/learning processes. Machine learning methods, such as meta-learning and multi-task learning, have been extended to modelling in federated infrastructure, but they have limitations in terms of scalability, robustness, and automation.

- **Protection of privacy** – the federated learning approach benefits privacy by keeping raw data (and possibly derived data) on each edge node/device in the federated network. However, sharing model updates in the network during the training/learning processes can expose sensitive information to a third party. Modular and differential approaches can enhance privacy in a federated infrastructure, but there may be trade-offs between privacy and model accuracy.

1.9 Discussion and Conclusions

LoRa is a wireless communication technology used in low-power, long-range communication applications. It provides low data rates to meet the requirements of remote edge nodes, which periodically send small amounts of sensor data. The architecture of LoRaWAN builds on a star topology that creates a single hop between an edge node (sensor IoT node) and the gateway. LoRa mesh networks are available for various applications that cannot be sufficiently managed by LoRaWAN architecture. The work in [32] has demonstrated that LoRaWAN applications can be extended using multi-hop LoRa, in which intermediate nodes can operate as repeaters that broadcast traffic to other LoRa nodes to reach a gateway.

The advantage of a LoRa mesh network is that the network coverage area can be expanded without adding more base stations. Furthermore, mesh networks combined with LoRa technology and AI-based techniques for routing optimisation can bring advantages to the application of the wireless sensor network in terms of improving the coverage area and promoting low power consumption.

Different wireless technologies, such as ZigBee, Z-Wave, BLE, Wi-SUN and Wi-Fi, use mesh topologies in which each device can be a router relaying the packet of the other devices to the end node. The main difference between LoRa and these other technologies is the ability for long-range transmission. This advantage can assist in expanding the network model without the need for additional base stations. In addition, low bandwidth makes LoRa resistant to channel noise, long-term relative frequency drift, Doppler effects and fading.

The key parameters used to configure the LoRa radio module are the modulation method, frequency range, bandwidth (BW), spreading factor (SF), coding rate (CR) and transmission power (TP). Artificial intelligence-based ML methods applied to LoRa and LoRaWAN for efficient resource management (e.g., BW, SF, CR and TP) can enhance LoRaWAN network performance and efficiency.

Edge AI solutions can be used in the processing modules of IoT devices that transmit information packets via the LoRa mesh network.

As the communication bandwidth of a LoRa link is low, performing ML on the IoT device allows for sending classification results rather than sending a higher amount of raw sensor data for remote classification. This saves the bandwidth of the low-capacity LoRa communication link.

Communication in a LoRa mesh network must adopt bandwidth-saving strategies, considering the duty cycle limitations of sub-GHz LoRa. Long range lacks packet delivery guarantees; for instance, using federated learning will require additional protocols for reliable messaging.

Depending on the setup and operational conditions, messages in a LoRa mesh network are also delivered with delays, excluding applications with strict real-time requirements. Consequently, distributed intelligence within a LoRa mesh network must determine the trade-off between using local computation and using communication resources. In these cases, there is a need for network integration of the LoRa mesh layer with the internet in full-stack edge IoT applications.

Long-range mesh networks and ML techniques deployed on edge IoT nodes can become communication substrates for building distributed intelligence with tiny edge nodes. The application can be extended using federated ML over LoRa communication, which is performed by embedded devices at the IoT layer.

Long-range mesh topology combined with intelligent gateways, AI-based routing optimisation and ML algorithms implemented in the processing nodes can be used for applications, such as intelligent lighting systems that provide extended coverage with limited data rates. Compared with other protocols for controlling large numbers of light devices, this technology can be suitable for lighting control.

In addition, the presented technologies can offer several benefits that contribute to the improved efficiency, scalability, and reliability of agricultural sensor systems. Sensors placed further from the central control point can still communicate through intermediate nodes, extending the coverage range of the overall network.

Mesh networks are self-healing, meaning that if one sensor node fails or is disrupted, the network can dynamically reroute data through alternative paths. This is an important feature, for example, in agriculture, in which environmental factors, such as weather, crop growth and equipment malfunctions, can temporarily disrupt communication. Sensor nodes equipped with edge AI functionality can detect or even predict such situations to improve system reliability and maintainability and, as a result, reduce costs.

Mesh networks can easily accommodate the addition of new sensor nodes without requiring significant infrastructure changes. This scalability is crucial, for example, in agriculture, in which the number of sensors might need to increase as the plantation expands or as new monitoring needs arise.

Communication through intermediate nodes alleviates the requirement for additional infrastructure components, thereby decreasing overall system costs. Moreover, agricultural sensor nodes frequently function in areas that are remote or difficult to reach, underscoring the importance of battery life. Leveraging low-power mesh protocols enables sensors to operate for extended durations without frequent battery replacements.

Acknowledgements

This research was conducted as part of the EdgeAI "Edge AI Technologies for Optimised Performance Embedded Processing" project, which has received funding from KDT JU under grant agreement No 101097300. The KDT JU receives support from the European Union's Horizon Europe research and innovation program and Austria, Belgium, France, Greece, Italy, Latvia, Luxembourg, Netherlands, and Norway.

The authors thank Fetze Pijlman and Jean-Paul Linnartz from Signify and the Eindhoven University of Technology, the Netherlands, for all their careful, constructive, and insightful comments and feedback about this work.

References

[1] LoRa (PHY). SEMTECH. Available at: https://www.semtech.com/lora/what-is-lora

[2] LoRa Alliance. https://lora-alliance.org/

[3] Semtech. Semtech SX128x Long Range Datasheet. 2019. Available online: https://semtech.my.salesforce.com/sfc/p/#E0000000JelG/a/2R000000HVET/HfcgiChyabtiPTh6EjcDM6ZEwAOQV7IirEmRULgggMM

[4] Semtech. Application Note: Ranging with the SX1280 Transceiver. 2017. Available online: https://semtech.my.salesforce.com/sfc/p/#E0000000JelG/a/2R000000HVET/HfcgiChyabtiPTh6EjcDM6ZEwAOQV7IirEmRULgggMM

[5] OpenThread. Available online: https://openthread.io/

[6] Thread. Available online: Available online: https://www.threadgroup.org/

[7] ZigBee Mesh Network. Available online: https://www.emcu-homeautomation.org/zigbee-mesh-network-ver-3-introduction/

[8] Zigbee. The Full-Stack Solution for All Smart Devices. Connectivity Standards Alliance. Available online: https://csa-iot.org/all-solutions/zigbee/

[9] G. R. Hiertz et al., "IEEE 802.11s: The WLAN Mesh Standard," in IEEE Wireless Communications, vol. 17, no. 1, pp. 104-111, February 2010, doi: 10.1109/MWC.2010.5416357.

[10] B.A.T.M.A.N. protocol concept. Available online: https://www.open-m esh.org/projects/open-mesh/wiki/BATMANConcept

[11] D. Johnson, N. Ntlatlapa, and C. Aichele, Simple pragmatic approach to mesh routing using BATMAN. 2nd IFIP International Symposium on Wireless Communications and Information Technology in Developing Countries, CSIR, Pretoria, South Africa, 6-7 October, pp 10, 2008. Available at: http://hdl.handle.net/10204/3035

[12] S.M. Darroudi, and C. Gomez, Bluetooth low energy mesh networks: A survey. Sensors, 17(7), p.1467, 2017. Available at: https://doi.org/10.3 390/s17071467

[13] R. Heydon, and N. Hunn, Bluetooth low energy. In CSR Presentation, Bluetooth SIG. 2012. Available at: https://www.Bluetooth.Org/DocM an/handlers/DownloadDoc.Ashx

[14] H. S. Kim, S. Kumar, and D. E. Culler, "Thread/Open thread: A compromise in low-power wireless multihop network architecture for the internet of things," IEEE Communications Magazine, vol.57, no.7, pp.55-61, 2019. Available at: https://par.nsf.gov/servlets/purl/10136090

[15] S.C. Ergen, ZigBee/IEEE 802.15. 4 Summary. UC Berkeley, September, 10(17), p.11, 2004. Available at: http://users.eecs.northwestern.edu/~p eters/references/ZigtbeeIEEE802.pdf

[16] P. Lech, P. Włodarski, Analysis of the IoT WiFi Mesh Network. In: R. Silhavy, R. Senkerik, Z. Kominkova, Z. Oplatkova, Z. Prokopova, P. Silhavy, (eds) Cybernetics and Mathematics Applications in Intelligent Systems. CSOC 2017. Advances in Intelligent Systems and Computing, vol 574. Springer, Cham, 2017. Available at: https://doi.org/10.1007/97 8-3-319-57264-2_28

[17] W. S. Conner "IEEE 802.11 TGs Functional Requirements and Scope," IEEE802.11- 04/1174r13, Jan. 2005.

[18] H. Harada, K. Mizutani, J. Fujiwara, K. Mochizuki, K. Obata, and R. Okumura, IEEE 802.15. 4g based Wi-SUN communication systems. IEICE Transactions on Communications, 100(7), pp.1032-1043, 2017. Available at: https://www.jstage.jst.go.jp/article/transcom/E100.B/7/E 100.B_2016SCI0002/_pdf/-char/en

[19] S. M. Hassan, R. Ibrahim, K. Bingi, T. D. Chung, N. Saad, Application of Wireless Technology for Control: A WirelessHART Perspective, Procedia Computer Science, Volume 105, pp. 240-247, 2017. ISSN 1877-0509. Available at: https://doi.org/10.1016/j.procs.2017.01.217

[20] J. Song, S. Han, A., Mok, D. Chen, M. Lucas, M. Nixon, and W. Pratt, WirelessHART: Applying wireless technology in real-time industrial

process control. In 2008 IEEE Real-Time and Embedded Technology and Applications Symposium, pp. 377-386, April 2008. Available at: https://www.cecs.uci.edu/~papers/cpsweek08/papers/rtas08/9B.pdf

[21] A. Saifullah, Y. Xu, C. Lu and Y. Chen, "Real-Time Scheduling for WirelessHART Networks," 2010 31st IEEE Real-Time Systems Symposium, San Diego, CA, USA, 2010, pp. 150-159. Available at: https://doi.org/10.1109/RTSS.2010.41

[22] T. Lennvall, S. Svensson and F. Hekland, "A comparison of WirelessHART and ZigBee for industrial applications," 2008 IEEE International Workshop on Factory Communication Systems, Dresden, Germany, 2008, pp. 85-88. Available at: https://doi.org/10.1109/WFCS.2008.4638746

[23] P.A.M. Devan, F.A. Hussin, R. Ibrahim, K. Bingi, F. A. Khanday, A Survey on the Application of WirelessHART for Industrial Process Monitoring and Control. Sensors 2021, 21, 4951. Available at: https://doi.org/10.3390/s21154951

[24] S. J. Danbatta and A. Varol, "Comparison of Zigbee, Z-Wave, Wi-Fi, and Bluetooth Wireless Technologies Used in Home Automation," 2019 7th International Symposium on Digital Forensics and Security (ISDFS), Barcelos, Portugal, 2019, pp. 1-5. Available at: https://doi.org/10.1109/ISDFS.2019.8757472

[25] M. Lilli, C. Braghin, and E. Riccobene, Formal Proof of a Vulnerability in Z-Wave IoT Protocol. In Proceedings of the 18th International Conference on Security and Cryptography (SECRYPT 2021), pages 198-209, 2021. ISBN: 978-989-758-524-1. doi: 10.5220/0010553301980209. Available at: https://www.scitepress.org/Papers/2021/105533/105533.pdf

[26] Z-Wave Alliance. Available at: https://z-wavealliance.org/

[27] J. Olsson, 6LoWPAN demystified. Available at: https://www.ti.com/lit/wp/swry013/swry013.pdf?ts=1691185461616&ref_url=https%253A%252F%252Fwww.google.com%252F

[28] G. Mulligan, The 6LoWPAN architecture. EmNets '07: Proceedings of the 4th workshop on Embedded networked sensors, June 2007, pp. 78–82, 2007. Available at: https://doi.org/10.1145/1278972.1278992

[29] A. Tønnesen, Implementing and Extending the Optimized Link State Routing Protocol, 2004. Available at: http://www.olsr.org/docs/report.pdf

[30] T. Clausen, and P. Jacquet, Optimized Link State Routing Protocol (OLSR), Oktober 2003. Available at: https://datatracker.ietf.org/doc/html/rfc3626

[31] P. Jacquet, P. Muhlethaler, T. Clausen, A. Laouiti, A. Qayyum, and L. Viennot, Optimized link state routing protocol for ad hoc networks, 2001. Available at: https://ieeexplore.ieee.org/document/995315

[32] J.R. Cotrim, and J.H. Kleinschmidt, "LoRaWAN mesh networks: A review and classification of multihop communication", Sensors, 20 (15) (2020). Available at: https://www.mdpi.com/1424-8220/20/15/4273

[33] J. Wang, M. Abolhasan, D. R. Franklin, and F. Safaei, OLSR-R∧3: Optimised link state routing with reactive route recovery, 2009. Available at: https://ro.uow.edu.au/cgi/viewcontent.cgi?referer=&httpsredir=1&article=1791&context=infopapers

[34] C.E. Perkins and E.M. Royer, Ad hoc On Demand Distance Vector (AODV) Routing, 1999. Available at: https://ebelding.cs.ucsb.edu/sites/default/files/publications/wmcsa99.pdf

[35] C. Perkins, E. Belding-Royer, and S. Das, Ad hoc On-Demand Distance Vector (AODV) Routing, July 2003. Available at: https://datatracker.ietf.org/doc/html/rfc3561

[36] D. Johnson, D. Maltz, and J. Broch, DSR: The Dynamic Source Routing Protocol for Multi-Hop Wireless Ad Hoc Networks. Available at: https://cs.brown.edu/courses/cs295-1/dsr-chapter00.pdf

[37] D. Johnson, Y. Hu, and D. Maltz, The Dynamic Source Routing Protocol (DSR) for Mobile Ad Hoc Networks for IPv4, 2007. Available at: https://datatracker.ietf.org/doc/html/rfc4728

[38] O. Iova, G. P. Picco, T. Istomin, and C. Kiraly, RPL, the Routing Standard for the Internet of Things . . . Or Is It?, 2017, Available at: https://hal.science/hal-01647152/document

[39] T. Tsvetko. RPL: IPv6 routing protocol for low power and lossy networks. Sensor nodes–operation, network and application (SN), 2011, 59. Jg., Nr. 2. Available at: https://citeseerx.ist.psu.edu/document?repid=rep1&type=pdf&doi=59b65811b94ba2162a9083744aef83fe09d381b0#page=67

[40] T. Winter, and P. Thubert, RPL: IPv6 routing protocol for low-power and lossy networks, 2012. Available at: https://datatracker.ietf.org/doc/html/rfc6550

[41] Li, T., Sahu, A. K., Talwalker, A., and Smith, V. "Federated Learning: Challenges, Methods, and Future Directions," 2019. Available at: https://arxiv.org/pdf/1908.07873.pdf

[42] Iqbal, Z. and Chan, H.Y. "Concepts, Key Challenges and Open Problems of federated learning," International Journal of Engineering, 2021. Available at: https://www.ije.ir/article_132537.html

[43] Almanifi, O. R. A., Chow, C-O., Tham, M-L., Chuah, J. H.m and Kanesan, J. "Communication and computation efficiency in federated Learning: A survey," Elsevier, ScienceDirect, Internet of Things, Volume 22, 2023. Available at: https://www.sciencedirect.com/science/article/pii/S2542660523000653

2

Edge AI Lifecycle Management

Dinu Purice[1], Francesco Barchi[2], Thorsten Röder[1], and Claus Lenz[1]

[1]Cognition Factory GmbH, Germany
[2]Universita di Bologna, Italy

Abstract

This chapter aims to define and interpret phases of the AI Lifecycle for Edge AI applications. We highlight common pitfalls that can arise when developing and maintaining AI models at the edge and outline best practices that are recognized in academia and industry, with the goal of developing a well-established taxonomy and pipeline for the lifecycle of Edge AI. We lay out that edge-based AI is seen as a natural extension of the cloud-based AI paradigm, solving problems related to real-time responsiveness, privacy, and independent operation closer to the source of the data. The challenges of edge-use cases are summarized, including limited network access, limited computational resources, and the need for customised deployment and maintenance procedures.

Keywords: machine learning (ML), software development lifecycle (SDLC), system-on-a-chip (SoC), deep learning (DL), dataset curation, edge AI, cloud-centric AI, model compression, deployment, monitoring, continuous learning, edge hardware.

2.1 Introduction and Background

In an ever more digital world, AI-based solutions have proven to be a driving force that is reshaping industries at an unprecedented pace. As artificial neural network architectures are growing, cloud-centred AI is a feasible

DOI: 10.1201/9788770041027-2   43

approach to deal with increasing computational requirements as no dedicated hardware is required and cloud-based systems can scale with application demands. This also unlocked the potential of AI-based solutions in industrial applications, followed by a continuous adoption of such technologies. Certain applications such as autonomous driving, financial trading, and healthcare monitoring. impose strict latency and availability requirements, which, coupled with concerns of data privacy and bandwidth availability, result in a set of requirements that cloud computing is unable to satisfy. Incorporating local data processing can be the key to achieving fast response and real-time latency, decoupled from the inherent delays arising from device-to-cloud communication. It enables decentralized solutions capable of inferring off-line, increasing service availability while lowering bandwidth and power requirements. In addition, it improves data privacy. By delegating computation closer to the edge, relevant features can be extracted, and private ones obfuscated before any data gets transferred over a network. This also reduces the size of data transfers, further easing the requirements on the network infrastructure of the overall solution.

EdgeAI refers to the practice of doing AI computations near the users at the networks edge instead of centralised locations [1], and in this context, becomes a natural extension of the cloud-centric paradigm, enabling the transfer of computations closer to the data acquisition source [2]. The EdgeAI computing market was estimated at $9 *bn* in 2020, with projections to reach $59.6 *bn* by 2030 [3]. Following this definition, the term edge device can describe both computation nodes between the edge and the Cloud (referred to as *fog computing*)[4], and lightweight processing units coupled with the sensors acquiring the data. The second type, further referred to as *low-powered devices*, includes field-programable gate arrays (FPGAs), tensor processing units (TPUs), media processing engines (MPE), as well as other processing units.

The main areas of applications of EdgeAI are among security, mobile networks, healthcare, voice and image analysis [5]. The list of tools and devices is constantly expanding as use-cases including predictive maintenance in industrial environments: sensors for predicting asset deprecation and maintenance timeline of production chains. Developments and innovations in the field of Edge AI happen both for software and hardware hand in hand, driven by the need for specialised frameworks for low-powered devices that cannot make use of containers and virtualisation typical of Cloud-based solutions. This enables a large variety of available solutions, of diverse complexity and computational power. On the other hand, the migration of existing *Machine*

Learning (ML) algorithms to the edge faces significant challenges due to the limited hardware capabilities associated with low-powered devices.

Various techniques of compression, enabling coping with the reduced hardware capabilities are an active subject of research. Most notable among compression techniques are *pruning-based* [6], which decreases the size and complexity of trained model by eliminating non-contributing components (weight, neuron, channel, filter) with minimal impact on accuracy and *quantization-based* [7], which reduce inference complexity by switching from the standard float-32 representation to more bit-conservative ones.

Within this chapter we define the stages of the Edge AI Lifecycle by augmenting the well-established *Software Development Life Cycle (SDLC)* with ML and edge-specific processes and stages.

For convenience, all phases are grouped into three stages: (I) Pre-Development, (II) Development, and (III) Production. It should be noted that, like the SDLC case, phases in the Edge AI Lifecycle can overlap and cycle back. An overview of the flow of the Edge AI Lifecycle is presented in Figure 2.1 with examples of frameworks used at each stage.

Starting with the first phase in the Lifecycle of any software solution is the *requirement formulation* phase, which includes functional and non-functional requirements. Based on these, we introduce the *ML methodology planning* phase, which includes data planning (describing the type of data and availability of labels to be used for training, validation, testing) as well as the choice of software frameworks and ML methodologies. A difference between the Cloud-centric approach and the edge approach, is the addition of a third component in the ML planning phase, namely the selection of inference hardware. Typically, in the case of Cloud-based solutions, the developed solution enters the deployment phase completely virtualized, and able to be deployed on any of the typical Cloud hardware, customisable within a few clicks on established platforms such as AWS. This is not the case for edge solutions, when the choice of hardware imposes restrictions on the software frameworks to be used, and on data formats. In the Edge use-case this step contains three intertwined components which impose limitations on each other. Following this phase, along with the *dataset assembly* phase, is the *Development* stage, consisting of the training, validation, evaluation, and optimisation phases. The conceptual difference between validation and evaluation is that validation chooses the best performing model of the many trained, while evaluation is used to obtain a representative estimate of its performance on unseen data. Following that, the optimisation stage then simplifies the obtained model while ensuring the accuracy does not drop below the specified requirements,

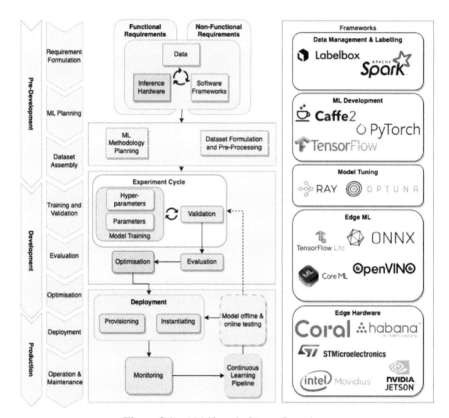

Figure 2.1   AI Lifecycle Stages Overview.

and depending on the techniques used can sometimes partially overlap with the training phase. Finally, the *Production* stage encompasses the deployment, operation, and maintenance phases. The following subchapters address each of the stages defined above and elaborate on the good practices and common pitfalls recognized within academic and industrial environments, with the goal of pushing towards an openly standardized approach to Edge AI development and deployment.

2.2 Pre-development

We define the Pre-Development stage as encompassing the **definition** and **planning** of the ML solution and associated hardware for inference, along with the assembly of the corresponding dataset.

The *definition phase* includes problem formulation, which represents the process of translating the real-world problem into a format that can be solved by a machine. For the *planning phase*, we introduced three intertwined categories in the previous sub-chapter: hardware, software, and data. To better understand the interaction between the three categories, we depart from the relationship between data types and machine learning algorithms capable of processing each type. Based on the problem statement, the type of data and the scope of the ML algorithms, several learning paradigms can be distinguished, as outlined in Table 2.1.

Each type of learning is equipped to handle different types of tasks, with their own requirements in terms of data and annotations. Unsupervised learning for example, being used mostly for extracting insights from large datasets with no labels, is rarely deployed to the edge. Typical edge use-cases refer instead to supervised (or semi-supervised, depending on the availability of labels for the data) or reinforcement learning. Every type of learning can in turn be further decomposed into different types of tasks. For example, a classification task can be formulated as binary classification, multiclass classification, or multi-label classification. A segmentation problem, very common in computer vision tasks, can be treated either as semantic segmentation (pixel-wise segmentation into foreground and background), or instance segmentation (different objects of the same class receiving distinct labels of the same class). Different formulations entail different labelling effort requirements. For example, although instance segmentation outputs

Table 2.1 Types of Learning and corresponding tasks

Type of Learning	Explanation	Application tasks
Supervised	Learning a function that maps an input to an output based on sample input-output pairs (labelled data)	Classification, Regression, Semantic Segmentation, Instance Segmentation
Unsupervised	Analyses unlabelled datasets without the need for human labelling (data-driven)	Feature Extraction, Trend identification, Clustering, Principal Component Analysis
Semi-supervised	Represents a combination of the above two types, typically used in dealing with a partially labelled dataset	Can be used to tackle both supervised and unsupervised type tasks
Reinforcement	Attempts to evaluate the optimal behaviour in a particular context or environment, based on reward or penalty.	Robotics, Autonomous Driving, Natural Language Processing

a detailed mask covering all pixels belonging to an instance of the object of the given class, in some applications a separation between foreground and background would suffice, significantly cutting the time required for labelling. It is recommended to choose the simplest problem formulation type which satisfies the task requirements, with the goal of minimising the resource requirement to prepare the dataset.

Dataset Formulation

The goal of the dataset preparation phase to create a set of data that is representative of the intended use-case, with sufficient examples to provide the developed neural network model with enough space during training for generalisation and identification of relevant features. It represents a critical stage which might make the production of a high accuracy model an impossible task, or more difficult than it needs to be. Hence, the right *domain knowledge* is required. Domain knowledge refers to the general background knowledge of the field or environment from which the data originates. It is particularly important for identifying outliers and non-representative data-points, detecting biases, and proposing attribute sampling methods, to reduce the non-informative data in the set. Equally important, domain knowledge is required for the formulation of data labelling guidelines, which help to ensure that the dataset is consistently annotated even if the annotation process is done by multiple experts as annotation variability must be kept to a minimum. Furthermore, to better combat the possibility of human error, data labelling by multiple experts in parallel can be an effective, albeit costly, solution. One common pitfall arising from insufficient data analysis and lack of domain knowledge during the pre-development stage is *concept drift* [8]. It refers to unforeseen changes in the relation between input and output data that are left unaccounted for. An example of concept drift is the shift in relation that might occur due to seasonal conditions e.g. summer to winter. Based on the nature of the change of the statistical properties of the predicted variable, the drift can be *sudden*, *gradual*, *incremental*, or *periodic*. The dangers of concept drift are amplified by the fact that its negative impacts on the accuracy are not detectable during training, and only become apparent during production, manifesting as degraded performance of the deployed solution. More on the detection and combating of concept drift is presented in the Production sub-chapter.

Data augmentation represents the process of "artificially" increasing a dataset by modifying copies of existing datapoints (*augmentation*) or synthetically generating new ones using the existing dataset (*synthetic*). Although typically used to expand datasets which have high costs of labelling, data augmentation techniques are also useful as an additional regularization factor, and to combat overfitting during training [9]. Another non-trivial use of data augmentation is to create datasets out of private data, when augmentation is used to obfuscate private features. Data augmentation can be counter-productive in cases with *data bias*, as the inherent bias in the data persists (and can be amplified) in the augmented dataset. Data bias describes the effect of over-representing certain elements in the dataset. It leads to models trained on it ending up "lazy", i.e. biased to predict the majority class.

There are multiple techniques of addressing the bias inherent in the data, at various stages of the AI Lifecycle. During pre-development, bias-compensating strategies include re-weighting and re-sampling the data, such that the dominating class becomes under-sampled. To further improve the generalisation capabilities and convergence of the developed model, it is recommended to make use of statistical rescaling techniques, such as *normalisation* and *standardisation*. Normalisation rescales the data to a [0,1] interval, and should be used when the distribution of the expected real-world data is unknown, while standardisation rescales the data such that the mean becomes zero and the standard deviation becomes one. It should be used when it can be assumed that the expected data follows a Gaussian distribution. Such techniques are helpful with improving the convergence speed during training, and with the regularisation of model weights. Before proceeding to Development, the dataset is split into training, validation, and test subsets. Most common split ratios include 60-80% for training, 10-20% for validation, and 10-20% for testing. While the train and validation subsets are actively used in the Development stage, the purpose of the test subset is to give an estimate of the performance of the resulting model on unseen data and should therefore only used for computing a final quality metric once the best performing model on the validation set is selected.

2.3 Development

In the domain of Edge AI, the development of lightweight neural network architectures has gained substantial importance. This is primarily driven by the increasing demand for precise and resource-efficient *Deep Neural Networks* (DNNs), especially in scenarios where these networks need to operate on resource-constrained edge computing devices. The development stage represents the most computationally intensive phase during which the network model is created, trained, evaluated, and optimised. At this stage in particular, comprehensive documentation is needed to record every step of model development, including weight initialisation and random number generator seed, to ensure the *reproducibility* and *transparency* of the process. Development is usually conducted in the Cloud or on-premise servers, where sufficient computational resources are available. This sub-chapter will provide a brief overview of state-of-the-art methodologies used for architecture design and training, as well as techniques for model compression and optimisation. Additionally, an overview of state-of-the-art hardware used for edge applications will be presented.

Model architecture development and training phase

Training phase represents the iterative process of exposing the neural network model to the dataset, enabling it to learn and adjusting its weights and biases, also known as model *parameters*, such that the accuracy of the model, measured on the train set (also referred to as fitting accuracy) increases. The process starts with model initialisation, during which the model *parameters* are either randomly initialised or pre-set in case of a pre-trained network, as well as with the selection of a *hyperparameter* set. The term hyperparameters refers to a broad set of choices made prior to the network training phase, and include design decisions of the network architecture (number of layers, neurons, filters, etc), learning rate, activation functions, optimisation algorithm, etc. The difference between model parameters and hyperparameters is that the first refers to the weights of the model trained through backpropagation applied on the model's loss function, while hyperparameters refer to top-level parameters controlling the learning process. Picking the right hyperparameters is not a straightforward process, and a sub-optimal choice would negatively influence the convergence of model training, as well as the resulting overall accuracy. The activity to identify suitable hyperparameters for DNN models within reasonable timeframes for novel applications has necessitated the adoption of automated pipelines. Trivial techniques for hyperparameter optimisation include *manual search*, *grid search*, and *random search*. These involve the launch of multiple experiments (i.e. independent training processes) with manually, grid-based, or randomly selected hyperparameters out of the set of possible values, which are then tried either sequentially or in parallel. The efficacy of each is then evaluated based on the performance of the model on the validation dataset, during or after training. Such methods do not guarantee that the optimal solution is found and are expensive in terms of computational resources and time. Due to the sheer complexity of manually exploring an extensive array of hyperparameter combinations, there has been a growing need for derivative network architecture search technologies. To this extent, more informed searching methods have been developed, such as *Evolutionary* and *Population-based Optimisation* [10]. These methods are adaptive, meaning they stop experiments in which the choice of hyper-parameters has proven to be sub-optimal, as measured by a user-defined fitness function. The terminated experiments are then replaced by new instances with hyperparameter sets derived from the more promising experiments. Such approaches are very efficient at minimising the training time and the hardware resources consumed compared to the previously mentioned classical search methods, and in addition provide a more exhaustive search over the hyperparameter space. Frameworks like *Ray Tune*, *Optuna*, and *Hyperopt* provide implementations for hyperparameter optimisation, and are compatible with most common ML frameworks such as *PyTorch*, *TensorFlow*, and *Keras*.

Another approach to hyperparameter optimisation is given by *Neural Architecture Search* (NAS) algorithms, which exhibit the capacity to optimize a diverse range of functions, encompassing both precision and complexity considerations, within a discrete search space. These algorithms have a considerable drawback due to the challenging evaluation step. Indeed, evaluating a sampled DNN necessitates a computationally intensive full training process.

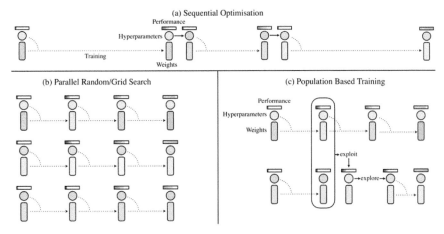

Figure 2.2 Overview of hyper-parameter training methodologies [10] illustrating (a) sequential optimisation; (b) parallel optimisation; (c) adaptive optimisation.

To alleviate this computational load, different techniques have been developed, such as the usage of reduced datasets, look-up tables and approximation of models to estimate cost-related metrics (memory occupancy, latency, energy consumption). *Differential Neural Architecture Search* (DNAS) represents a pivotal advancement in the realm of NAS, markedly reducing the time required for optimization. This reduction is achieved by transitioning from a discrete search space to a continuous one, rendering the problem addressable using gradient-descent optimization techniques. The central idea of DNAS revolves around the definition of a set of architectural parameters able to encode the selection of a DNN architecture from the search space. DNAS jointly optimizes these architectural parameters alongside the weights of the neural networks. This amalgamation of architectural parameter optimization and weight training within a continuous search space contributes to the accelerated optimization of DNN architectures, making DNAS a promising approach to exploring efficient and effective neural network design.

"[11] introduces DARTS Differentiable Architecture Search", addressing the challenges associated with scalability in architecture search. DARTS introduces the DNAS concept, framing architecture search as a differentiable problem. Through the continuous relaxation of architectural representations, DARTS enables accelerated search processes employing gradient descent techniques, significantly reducing search time. Extensive experiments have been conducted on diverse datasets, including CIFAR-10 and ImageNet, showing DARTS exceptional ability to uncover high-performance convolutional and recurrent architectures tailored specifically for image classification and language modelling tasks. This goal is especially relevant in a domain where optimized network architectures, capable of accommodating the constraints of edge devices, hold considerable importance, thus contributing to the advancement of EdgeAI model development. In [12], the researchers acknowledge the escalating demand for DNN models that strike a balance between precision and operational efficiency, a requirement in the context of edge computing. PLiNIO, is an open-source library that consolidates a comprehensive set of cutting-edge DNN design automation techniques into a user-friendly interface. These techniques, rooted in lightweight gradient-based optimization, simplify the intricacies of DNN development for edge applications. Through empirical assessments conducted on tasks pertinent to edge computing, the study demonstrates that PLiNIO yields many DNN solutions that surpass baseline models in respect of the delicate trade-off between accuracy and model size. It is worth noting that PLiNIO exhibits remarkable memory reductions, up to 94.34%, while maintaining accuracy levels, underscoring its pivotal role in EdgeAI model development.

In summary, derivative network architecture search technology, exemplified by pioneering frameworks such as DARTS, is pivotal in the EdgeAI model development. These innovative approaches make the optimization process more efficient, allowing us to navigate the complex landscape of hyper-parameter configurations and unveil DNN architectures that achieve optimal equilibrium between accuracy and model size. This research direction holds great promise for the future of Edge AI, where resource-efficient, high-performing neural network architectures serve as the bedrock for a wide range of applications.

Model validation phase

Model validation asseses the quality of the training process by measuring the accuracy of the model on a dedicated validation dataset (validation accuracy). It goes hand in hand with the training phase. Insights acquired from the validation accuracy assessment are then used to compare different training instances to identify the optimal hyperparameter choices, and to assess when the training process should be stopped. Typically, an *early stopping* mechanism is used for this purpose, which monitors the development of validation accuracy and stops the training once the accuracy reaches a plateau or starts degrading. Failing to stop a training session in time is one of the causes of *overfitting*, occurring when the model learns patterns unique to the training set that do not apply to real-world data. An overfitted model is typified by a high discrepancy between the fitting and validation accuracies and performs poorly on unseen data. Various techniques to combat the overfitting effect exist and can be grouped by mechanism as presented in Table 2.2.

Model evaluation phase

The evaluation phase starts once the training has been completed and the best performing instance of the model has been identified. The goal of this phase is to assess how well the trained model generalises to new, unseen data, thus emulating a real-world scenario. The accuracy of the model on the test dataset is measured, and serves as a final, unbiased indicator of the model's

Table 2.2   Techniques to combat overfitting.

Mechanism type	Technique description
Data-based	**More training data** – most straightforward, increase the diversity of the training data by adding additional datapoints
	Data augmentation – artificially increase the diversity of the training data through augmentation techniques and the addition of noise.
	K-fold cross validation – split the dataset into K subsets, with each subset used for validation set once while the others are used for training. Other similar cross validation techniques include stratified cross-validation, leave-one-out-cross-validation (LOOCV), etc.
Regularisation-based	**L1 and L2 Regularisation** – penalise complex model weights by adding their L1 or L2 norm to the loss function as an additional term
	Dropout [13] – randomly deactivate a fraction of neural network neurons during each training iteration
Feature-based	**Feature engineering** – manual selection, transformation, and creation of features from the original data
	Pruning – removal of parameters from a network based on their usefulness to the inference output, thus reducing the model's complexity
Inference-timed	**Model ensembling** [14] – combine predictions from multiple models to produce a single optimal predictive model
Training-timed	**Early stopping** – stops the training process once the validation accuracy stops improving

performance. The measured evaluation accuracy must not then be used to make any further decisions about the model's architecture, hyperparameters, or any other aspect of training. Doing so would represent a form of *data leakage* when information from outside the training and validation phases makes its way into the training pipeline and undermines the validity and estimated evaluation accuracy of the trained model. Instead, in case the measured evaluation accuracy does not satisfy the requirements set in the previous stage, the whole development process must be restarted, with new dataset splits.

Model compression phase

Compressing a DNN model is crucial for making it more suitable for deployment on resource-constrained devices. There are several techniques available to achieve DNN model compression, as outlined in Table 2.3.

Table 2.3 Compression Techniques

Compression Technique	Technique description
Weight Quantisation	Involves representing the model's weights with a lower bit precision than the standard 32-bit floating-point numbers. Common bit-widths include 8-bit or even lower, reducing memory and computational requirements.
Model Quantisation	Involves quantizing activations during inference. This can further reduce memory and computation requirements by using lower-precision representations for intermediate activations.
Pruning	Involves removing unimportant/low-magnitude weights or neurons from the model. These elements contribute minimally to the model's performance, so their removal can significantly reduce model size and inference time without a significant loss in accuracy.
Knowledge Distillation	Represents training a smaller student model to mimic the behaviour of a larger, more complex teacher model. This transfer of knowledge from the teacher to the student model results in a smaller and more efficient model that maintains most of the teacher's accuracy.
Knowledge Pruning	This approach combines knowledge distillation with pruning. The teacher model is first pruned to a smaller size, and then a student model is trained to mimic the pruned teacher. This results in a more compact model while maintaining the knowledge of the original, larger model.
Low-Rank Factorization	This technique decomposes the weight matrices of the model into lower-rank matrices. By doing this, you can reduce the number of parameters in the model, leading to a smaller model with less computational overhead.
Sparse Models	Sparse models are models with a substantial number of zero-valued weights. Techniques like sparse training or structured sparsity constraints can be applied to encourage weight sparsity, resulting in a more compact model.
Compact Architectures	Using model architectures designed for efficiency, such as *MobileNet*, *EfficientNet*, or *SqueezeNet*, can lead to smaller models that maintain competitive performance on various tasks.
Transfer Learning	Instead of training a model from scratch, one can use pre-trained models as a starting point and fine-tune them to the specific task at hand. This approach leverages the knowledge learned from a larger dataset and model, resulting in a smaller model customised for the specific task.

These techniques can be used individually or in combination to achieve the desired level of compression while minimizing the impact on model accuracy and performance. The choice of technique(s) depends on the specific requirements of the task at hand and the available computational resources. Quantization, for example, has evolved significantly in the context of Edge AI applications. Its history is marked by the pursuit of approximating floating-point weights and activations with low bit-width integers, ultimately aimed at reducing model size and enhancing operational efficiency. Particularly at the edge, where computational resources are constrained, quantization is a critical factor in making DNNs more viable [15]. In the past, quantization was often applied post-training, essentially mapping the high-precision model to a lower-precision representation. However, a significant breakthrough came with the introduction of *Quantization-Aware Training* (QAT) [16]. QAT enables DNNs to adapt to the effects of quantization during the training process, mitigating the subsequent drop in accuracy that occurs with post-training quantization. Standard fixed-precision quantization assigns a uniform integer bit-width to the entire DNN, neglecting the unique sensitivity of each layer to precision reduction. Recognizing this limitation, the field advanced with mixed-precision methods [17]. These approaches introduce variability in bit-width assignment, quantizing different subsets of the DNN at varying levels of precision. This innovation, however, introduces a challenging optimization problem, demanding the identification of precise bit-width assignments that strike an optimal balance between model accuracy and computational complexity. The challenge grows exponentially with the number of considered bit-widths, making it a computationally intensive effort. Several mixed-precision strategies have emerged to address this complexity-accuracy trade-off, representing a parallel development orthogonal to NAS. Additionally, some strategies employ reinforcement learning techniques to automate bit-width assignment. Recently, a gradient-based method inspired by the principles of DNAS was introduced, enabling bit-width assignment during training [18]. This method dynamically quantizes data at various precisions and selects an optimal precision during the training process. In essence, quantization techniques have witnessed a historical shift from post-training conversion to in-training adaptation, reflecting the growing importance of model efficiency in the context of Edge AI applications and the innovative approaches developed to optimize this critical aspect of DNNs.

Hardware for Edge AI

Edge AI relies on a variety of hardware components and platforms to enable efficient and real-time inference. Many edge devices, such as smartphones, IoT devices, and embedded systems, use SoCs that integrate various components like CPU, GPU, DSP, and often hardware accelerators like *Neural Processing Units* (NPUs) or *Field-Programmable Gate Arrays* (FPGAs). These compact and power-efficient chips are well-suited for running AI workloads at the edge. General-purpose CPUs are still widely used in edge devices for AI inference, especially for less demanding tasks. Many modern CPUs come with support for hardware-based vectorization and optimizations like SIMD (Single Instruction, Multiple Data) instructions to accelerate AI workloads. GPUs, originally designed for graphics rendering, are highly parallel processors that excel at performing matrix operations essential for deep learning. Edge devices equipped with GPUs can leverage their computational power for AI tasks. Specialized NPUs designed explicitly for accelerating deep learning workloads are increasingly integrated into SoCs for edge devices and provide hardware acceleration for AI inference, improving both speed and energy efficiency, but generally have higher power consumption than dedicated hardware. FPGAs offer hardware programmability, making them adaptable to specific AI models and tasks. They are commonly used in scenarios where low latency and real-time processing are crucial, such as autonomous vehicles and robotics. AI-specific accelerators, like Google's Tensor Processing Unit (TPU) and Intel's Movidius VPU, are custom-designed chips optimized for AI workloads. These accelerators are highly efficient for tasks like image recognition, object detection, and voice processing, making them valuable for Edge AI applications with stringent requirements. Depending on the specific needs of an Edge AI application, custom hardware solutions may be developed to meet unique demands, such as specialized hardware for robotics. The choice of hardware for Edge AI depends on factors such as the specific AI workload, power constraints, latency requirements, and cost considerations. Many Edge AI applications use a combination of these hardware components to optimize performance, power efficiency, and resource utilization for AI inference. Of particular significance are the architectural advancements that have emerged in recent years, owing to the advent of RISC-V, an open Instruction Set Architecture (ISA) that empowers hardware developers to devise pioneering and high-performance solutions. As an exemplar, the GreenWaves GAP8 processor, equipped with eight CV32E40P cores [19], delivers 22.65 Giga Operations Per Second (GOPS) with an exceptional power efficiency of 4.24 milliwatts per GOP (mW/GOP). This technological achievement was effectively harnessed for the autonomous navigation of a micro-drone through the execution of a neural network [20].

2.4 Production

With the increase in maturity of Machine Learning algorithms, the issue of efficient deployment and maintenance comes more and more into focus. This has led to the emergence of the MLOps field, which handles the tasks of deployment, monitoring, and operations of ML models. Provisioning, the starting point of deployment to the edge, represents the translation of the model to the specific architecture of the hardware. Within the Cloud paradigm, provisioning is less critical, as the models are traditionally deployed through virtual machines and containers, isolated from underlying hardware. Deployment to the edge however, particularly to low-power devices, requires the use of specialized frameworks, often developed and maintained by the manufacturers of said devices. Typically said frameworks consist of an intermediate representation component, which represents the prepared model in a lightweight, optimised state, and an inference engine which runs the model. Intel's *OpenVINO* toolkit is one such example, best suited for Intel's CPUs, GPUs, as well as GNAs. *CoreML* is compatible with Apple devices, while *Tensorflow Lite* is best used for Android and Coral TPUs. At the same time attempts are made at creating universal formats, such as *ONNX*, which supports a variety of frameworks used for developing ML models, such as *Tensorflow*, *PyTorch* and *Caffe*, and make them available on various hardware. Depending on the requirements and complexity of the application, the model can then be deployed to function in an online or offline mode. For describing the monitoring phase, we will assume an online functioning mode, with at least occasional network connectivity to a Cloud-based managing framework. The data used for training does not always accurately reflect the real-world encountered by a deployed model in the long term, reflected in degrading model performance over time, adjustments must be made based on insights acquired during the production stage. Variations in production data distributions, a symptom of this effect, can be detected with data drift detection algorithms, such as the *Kolmogorov-Smirnov* test, *Population Stability* Index, *Page-Hinkley* method, etc. The phenomenon of taking such insights into account and modifying the deployed model based on them is called *Continuous Learning* and represents a technique of proactive intervention to combat model drift.

After sufficient new data are acquired during production, a data curation stage is triggered, in which the data is prepared for a fine-tuning session. The fine-tunning session mirrors the training and validation pipeline, followed by an offline testing stage determining if the resulting fine-tuned model has

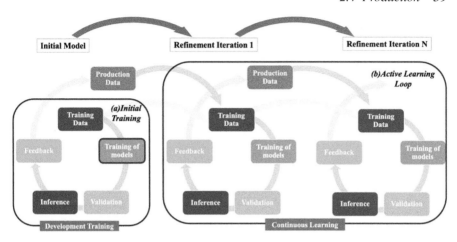

Figure 2.3 Model Training Overview illustrating (a) training during the development stage and (b) training during the production stage.

improved or downgraded its performance. Finally, the fine-tuned model is deployed in parallel to the production version, and their predictions compared in online testing, in which their comparative accuracies on unseen data are evaluated. In case the fine-tuned model is performing better, it takes the place of the previous version of the model, and the other is removed from service. Good version control is essential at this stage, to track model development and to keep the older versions as fall-back options, to be made available in case of unforeseen deviations by the active model. An overview of the ML procedures taking place within the continuous learning paradigm is presented in Figure 2.3. Here during stage (a) the optimal hyperparameters are found and the model is trained on the initial data, and in stage (b) the model makes use of the continuous learning pipeline to fine-tune its weights based on feedback from the production environment. It should be noted however that particularly in the case of Edge AI, where the production stage takes place on distributed, low-powered hardware, the infrastructure required to enable the continuous learning pipeline becomes more convoluted than in the Cloud-centric case. The issues that it needs to consider are the reduced computing power, which must be shared between the inferring component and the data acquisition component, and the limited bandwidth to be used for data transfer and model re-deployment, as well as the fact that new models must be deployed on each device. Overall, the infrastructure must support Edge-to-Cloud integration for transfers of fresh data, Model Version Control

Table 2.4   Types of Automation based on the definition by SAE International [21]

Level of Automation	Stage Name	Stage explanation
Level 0	No Automation	The human operator performs all tasks without input from the machine
Level 1	Assistance	Limited assistance is provided to the human operator in completing specific tasks
Level 2	Partial Automation	The machine takes over some of the task, but continuous human monitoring is required
Level 3	Conditional Automation	The machine can perform most tasks independently, but human intervention is required in case of complex and unexpended situations
Level 4	High Automation	The machine can perform most tasks independently, human intervention required in exceptional cases
Level 5	Full Automation	Human operator not needed at all. The machine can perform independently including in exceptional conditions

for tracking and updating models as needed, and Over-the-Air (OTA) updates for the deployed models.

Depending on the level of autonomy of the deployed AI solution, as well as the requirements of the human factor in inference monitoring, different levels of autonomy can be defined. Currently, there is a system in place for autonomy in vehicles developed by *SAE International* [21], which we will use as a starting point to generalise guidelines for the autonomy of Edge AI solutions, presented in Table 2.4.

The difficulties of reaching level three and above, as defined in the table above, particularly in the case of autonomous driving, lie in the unpredictability of the environment in which an autonomous vehicle operates. In case of controlled environments, as is usually the case for applications within factories and assembly lines, the probability of unexpected and exceptional cases diminishes considerably, easing the transition to high and full automation.

2.5 Conclusion

In this chapter we have presented Edge AI as a natural extension of the Cloud-centric AI paradigm that enables solutions for use-cases with strict latency and data privacy requirements. The challenges and novel research directions arising from the transition towards the edge are summarised, including the development of compression techniques aimed at reducing model complexity and inference time with minimised accuracy losses, as well as the design of compact, low-power hardware and associated software for network deployment. The standard SDLC is expanded to include the emergent set of good practices of ML development, deployment to the edge and maintenance, and encapsulated within the Edge AI Lifecycle. Divided into a pre-development, development, and production stage, common pitfalls and good practices are outlined, with the goal of pushing towards a well-established pipeline and taxonomy in the field of Edge AI. The Pre-Development section summarises the processes of dataset assembly, and problem definition with the translation of the task into one of the ML paradigms. Following that, the Development section addresses the emergent automation of the network design phase, as introduced by evolutionary hyper-parameter search algorithms, and NAS-based methodologies. The section goes on to describe model validation and evaluation tactics, common to all ML applications. The specifics of edge use-cases are then addressed by a categorisation of model compression techniques, and an overview of available edge hardware. Finally, the Production section details a collection of frameworks used for the deployment of optimised models on dedicated hardware and outlines the importance of production monitoring and continuous learning pipelines.

Acknowledgements

This research was conducted as part of the EdgeAI "EdgeAI Technologies for Optimised Performance Embedded Processing" project, which has received funding from KDT JU under grant agreement No 101097300. The KDT JU receives support from the European Union's Horizon Europe research and innovation program and Austria, Belgium, France, Greece, Italy, Latvia, Luxembourg, Netherlands, and Norway.

References

[1] R. Singh and S. S. Gill, "Edge AI: A survey," *Internet of Things and Cyber-Physical Systems*, vol. 3, pp. 71–92, Jan. 2023, doi: 10.1016/J.IOTCPS.2023.02.004.

[2] N. Kukreja *et al.*, "Training on the Edge: The why and the how," *Proceedings - 2019 IEEE 33rd International Parallel and Distributed Processing Symposium Workshops, IPDPSW 2019*, pp. 899–903, Feb. 2019, doi: 10.1109/IPDPSW.2019.00148.

[3] "AI Edge Computing Market Statistics | Industry Forecast - 2030." https://www.alliedmarketresearch.com/ai-edge-computing-market-A14885 (accessedAug.22,2023).

[4] S. Yi, Z. Hao, Z. Qin, and Q. Li, "Fog computing: Platform and applications," *Proceedings - 3rd Workshop on Hot Topics in Web Systems and Technologies, HotWeb 2015*, pp. 73–78, Jan. 2016, doi: 10.1109/HOTWEB.2015.22.

[5] T. Sipola, J. Alatalo, T. Kokkonen, and M. Rantonen, "Artificial Intelligence in the IoT Era: A Review of Edge AI Hardware and Software," *Conference of Open Innovation Association, FRUCT*, vol. 2022-April, pp. 320–331, 2022, doi: 10.23919/FRUCT54823.2022.9770931.

[6] D. Blalock, J. J. G. Ortiz, J. Frankle, and J. Guttag, "What is the State of Neural Network Pruning?," Mar. 2020, Accessed: Aug. 22, 2023. [Online]. Available: https://arxiv.org/abs/2003.03033v1

[7] I. Hubara, M. Courbariaux, D. Soudry, R. El-Yaniv, and Y. Bengio, "Quantized Neural Networks: Training Neural Networks with Low Precision Weights and Activations," *Journal of Machine Learning Research*, vol. 18, pp. 1–30, 2018.

[8] J. Lu, A. Liu, F. Dong, F. Gu, J. Gama, and G. Zhang, "Learning under Concept Drift: A Review," *IEEE Trans Knowl Data Eng*, vol. 31, no. 12, pp. 2346–2363, Apr. 2020, doi: 10.1109/TKDE.2018.2876857.

[9] C. Shorten and T. M. Khoshgoftaar, "A survey on Image Data Augmentation for Deep Learning," *J Big Data*, vol. 6, no. 1, pp. 1–48, Dec. 2019, doi: 10.1186/S40537-019-0197-0/FIGURES/33.

[10] M. Jaderberg *et al.*, "Population Based Training of Neural Networks," Nov. 2017, Accessed: Aug. 22, 2023. [Online]. Available: https://arxiv.org/abs/1711.09846v2

[11] H. Liu, K. Simonyan, and Y. Yang, "DARTS: Differentiable Architecture Search," *7th International Conference on Learning Representations, ICLR 2019*, Jun. 2018, Accessed: Sep. 07, 2023. [Online]. Available: https://arxiv.org/abs/1806.09055v2

[12] D. J. Pagliari, M. Risso, B. A. Motetti, and A. Burrello, "PLiNIO: A User-Friendly Library of Gradient-based Methods for Complexity-aware DNN Optimization," Jul. 2023, Accessed: Sep. 07, 2023. [Online]. Available: https://arxiv.org/abs/2307.09488v1

[13] N. Srivastava, G. Hinton, A. Krizhevsky, and R. Salakhutdinov, "Dropout: A Simple Way to Prevent Neural Networks from Overfitting," *Journal of Machine Learning Research*, vol. 15, pp. 1929–1958, 2014.

[14] M. A. Ganaie, M. Hu, A. K. Malik, M. Tanveer, and P. N. Suganthan, "Ensemble deep learning: A review," *Eng Appl Artif Intell*, vol. 115, p. 105151, Oct. 2022, doi: 10.1016/J.ENGAPPAI.2022.105151.

[15] R. Banner, Y. Nahshan, and D. Soudry, "Post-training 4-bit quantization of convolution networks for rapid-deployment," *Adv Neural Inf Process Syst*, vol. 32, Oct. 2018, Accessed: Sep. 07, 2023. [Online]. Available: https://arxiv.org/abs/1810.05723v3

[16] B. Jacob *et al.*, "Quantization and Training of Neural Networks for Efficient Integer-Arithmetic-Only Inference," *Proceedings of the IEEE Computer Society Conference on Computer Vision and Pattern Recognition*, pp. 2704–2713, Dec. 2017, doi: 10.1109/CVPR.2018.00286.

[17] Z. Dong, Z. Yao, A. Gholami, M. Mahoney, and K. Keutzer, "HAWQ: Hessian AWare Quantization of Neural Networks with Mixed-Precision," *Proceedings of the IEEE International Conference on Computer Vision*, vol. 2019-October, pp. 293–302, Apr. 2019, doi: 10.1109/ICCV.2019.00038.

[18] K. Wang, Z. Liu, Y. Lin, J. Lin, and S. Han, "HAQ: Hardware-Aware Automated Quantization with Mixed Precision," *Proceedings of the IEEE Computer Society Conference on Computer Vision and Pattern Recognition*, vol. 2019-June, pp. 8604–8612, Nov. 2018, doi: 10.1109/CVPR.2019.00881.

[19] A. Pullini, D. Rossi, I. Loi, G. Tagliavini, and L. Benini, "Mr.Wolf: An Energy-Precision Scalable Parallel Ultra Low Power SoC for IoT Edge Processing," *IEEE J Solid-State Circuits*, vol. 54, no. 7, pp. 1970–1981, Jul. 2019, doi: 10.1109/JSSC.2019.2912307.

[20] D. Palossi, A. Loquercio, F. Conti, E. Flamand, D. Scaramuzza, and L. Benini, "A 64mW DNN-based Visual Navigation Engine for Autonomous Nano-Drones," *IEEE Internet Things J*, vol. 6, no. 5, pp. 8357–8371, May 2018, doi: 10.1109/JIOT.2019.2917066.

[21] "SAE J3016 automated-driving graphic." https://www.sae.org/news/2 019/01/sae-updates-j3016-automated-driving-graphic(accessedSep. 01,2023).

3

Federated Learning: Privacy, Security and Hardware Perspectives

Taha Yassine Abidi, Iyad Dayoub, Elhadj Doguech, and Ihsen Alouani

Université Polytechnique Hauts-De-France, France

Abstract

Machine Learning (ML) models are being deployed in a wide range of domains owing to their capacity to deliver high performance across a range of challenging tasks including safety-critical and privacy-sensitive applications. Moreover, the computing requirements of increasingly complex ML models presents a significant challenge to the hardware industry.

Against this backdrop, Federated Learning (FL) has emerged as a promising technique that enables privacy-preserving development of ML models on low-energy Edge devices. FL is a distributed approach that enables learning from data belonging to multiple participants, without compromising privacy since user data are never directly shared. Instead, FL relies on training a global model by aggregating knowledge from local models. Despite its reputation as a privacy-enhancing strategy, recent studies reveal its susceptibility to sophisticated attacks that can undermine integrity and, as well as disrupt their operations. Notably, the constraints posed by the limited hardware resources in edge devices compound these challenges. Gaining insight into these potential risks and exploring hardware-friendly solutions is vital for effectively implementing trustworthy and power-efficient FL systems in edge environments.

DOI: 10.1201/9788770041027-3   65

This chapter contributes a review and perspective of the triad of privacy, security, and hardware optimization in FL settings.

Keywords: Federated Learning, Hardware Optimisation, ML Security, Privacy.

3.1 Introduction and Background

In this era of unprecedented data proliferation and exponential technological advancement, conventional centralized and cloud-based training and deployment of machine learning faces 2 main challenges:

- How to train and deploy accurate models in an energy-efficient and sustainable manner?
- How to guarantee the security and privacy of potentially sensitive data without compromising the learning process?

FL has emerged as a promising approach to address the challenges posed by decentralised data sources while preserving data privacy. Traditional centralised ML approaches require aggregating sensitive data from various sources into a central repository for training, which can raise concerns about data exposure and privacy. FL offers an innovative solution by enabling model training across multiple devices or data silos, without the need to centralise the data themselves. This distributed approach not only safeguards individual privacy but also optimises the utilisation of edge devices, edge servers, and cloud resources.

The key motivation behind FLis to leverage the collective intelligence of a network of devices while maintaining data locality. This is particularly crucial in scenarios where data is distributed across devices or locations, such as Internet of Things (IoT) ecosystems, healthcare networks, and financial institutions. By allowing devices to collaboratively learn a shared model while keeping their data local, federated learning can address challenges like network latency, bandwidth limitations, and data security.

In this chapter, we delve into the multifaceted aspects of FL, focusing on privacy, security, and the opportunities for hardware optimisation at the Edge. We explore the techniques that enable data privacy within FL, the security measures needed to protect against adversarial attacks, and the ways in which hardware constraints and advancements shape the landscape of FL. Through case studies and emerging trends, we aim to provide a comprehensive understanding of how federated learning empowers data-driven insight while upholding individual privacy, ensuring security, and harnessing the potential of diverse hardware resources.

This chapter not only sheds light on the current state of federated learning but also serves as a guide for researchers, practitioners, and policymakers who seek to navigate the intersection of machine learning, distributed systems, and data governance. As FL continues to evolve, it is imperative to appreciate its significance in reshaping the landscape of data-driven technologies, fostering collaboration, and advancing both technological and ethical dimensions in the digital era.

The structure of the chapter is crafted to offer a comprehensive exploration of the FL state-of-the-art. Our roadmap unfolds as follows: we begin with an initial introduction to the basics of FL and its applications, followed by an exploration of FL's constraints and limitations, including hardware resources, security, and privacy considerations. Finally, we conclude by underscoring the crucial requirement for balance among these varied aspects.

3.2 Federated Learning Overview

Training a deep neural network necessitates a significant amount of data, often representing the most valuable resource within a target environment: it can be of commercial value, be governed by privacy regulations, can be limited by user agreements (as illustrated by regulations such as HIPAA in the US and GDPR in Europe). In another scenario, data generated on Edge devices may face sharing restrictions due to privacy anxieties, bandwidth restrictions, or performance constraints.

FL recently emerged as a potential solution to the problems above. It enables participants to collaboratively train a federated model while preserving local data privacy. Within the FL framework, each participant trains a local model sharing it with a central server also known as a central aggregator. Data remain private to each participant. The server aggregates the local model updates into a single federated model and shares this model with the participants, creating an updated federated model that benefits from all the data without jeopardising its confidentiality. The model's refinement continues as participants deliver more updates.

FL encompasses three primary categories from a data partitioning perspective: horizontal FL, vertical FL and federated transfer learning [5]. This document, however, zooms on the most prevalent and widely used model, namely horizontal FL. In the subsequent section, we consider the intricacies

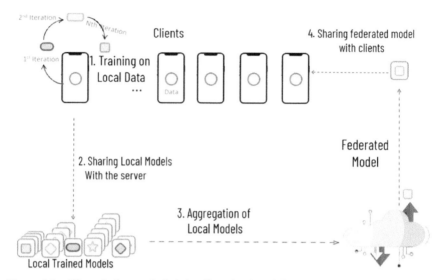

Figure 3.1   Client device sends their locally trained model updates to server for training the federated model.

of horizontal FL while also offering succinct insights into the other two models for context.

3.2.1 Horizontal Federated Learning

Participants train their local model with data that are in the same feature space. For example, two regional hospitals might contain different patient population data, with little to no intersection in the data (perhaps because the hospitals serve different regions). However, the activities of the two hospitals are similar with respect to each other and so their feature spaces are the same. During the training phase of the horizontal FL model, each of the participants trains its local model using the local, private, data and sends the gradients to the central aggregator. The central aggregator aggregates all of those local model updates to build a global shared model and return this back to all participants. Finally, each participant updates its local model using the result from the central aggregator.

3.2.2 Vertical Federated Learning

Vertical FL addresses the scenario where participants refine their respective local models using data samples derived from different feature spaces. For example, consider a hospital and a pharmacy in the same region. While there is likely a significant overlap in patient population data, the retained information (i.e., the features) for these patients vary due to the distinct functions of the two participants. For example, the hospital preserves the records of all users about their disease, diagnosis and information of treatment received while the pharmacy keeps the records of medicine purchasing history. Using a vertical FL system, the two institutions can collaboratively build a prediction model by aggregating those different features and calculating the gradients of their local data in a privacy preserving manner.

3.2.3 Federated Transfer Learning

Federated Transfer Learning [7] finds its niche in scenarios where users' datasets remain disjoint or share minimal overlap in both the samples and feature spaces. For instance, revisiting the hospital and pharmacy scenario recall that the feature spaces of their data have little overlap. If the two institutions are in different countries they would also have few, if any, common patients, making it impossible to apply VFL. FTL solves this problem by creating a common representation using transfer learning and using it to build a predictive model across the entire data set.

With a foundational understanding of FL in place, we now turn our attention to the challenges that accompany this paradigm. While FL offers a promising avenue for decentralised model training and data privacy preservation, it is essential to acknowledge its limitations.

3.3 Challenges and Limitations of Federated Learning

FL presents a trio of critical challenges that demand rigorous exploration: hardware resources, security, and privacy.

These dimensions shape the framework's efficacy and ethical underpinnings. In this section, we consider this interplay.

3.3.1 Security challenge

The distributed nature of FL, while preserving data locality, introduces complexities that require careful attention to ensure the confidentiality, integrity, and authenticity of the data and models being exchanged.

Adversarial attacks, stemming from both malicious clients and malicious servers, pose a significant threat to the security and integrity of FL by exploiting vulnerabilities inherent in the decentralised nature of the approach. These attacks aim to manipulate the training process and the resulting model's composition, leading to erroneous predictions and potential data exposure.

In this context, understanding the objectives that potential attackers pursue becomes crucial. These objectives can be categorised into three primary dimensions:

- Compromising System Integrity: Attackers aim to compromise the integrity of the FL system by tampering with the model's function. They induce misclassifications by poisoning individual local model updates or by colluding with other malicious participants.

- Compromising Data Confidentiality and Privacy: Adversaries target data confidentiality and privacy by attempting to infer private information or reconstruct original training samples. We will delve deeper into this topic in the upcoming section.

- Disrupting the Learning Process: Attackers seek to disrupt the learning process itself. This includes tactics such as initiating denial-of-service or impeding the convergence of the training process.

To achieve these objectives, adversaries deploy a range of strategic actions:

- Poisoning Attacks: Malicious actors maliciously alter either the training data or the model to corrupt the overall federated model's integrity. This compromise is executed with the intention of manipulating the model's behaviour to serve the attacker's motives.

- Privacy Attacks: Adversaries attempt to deduce sensitive information about the data, which will be discussed in detail in the subsequent section.

- Disruption Attacks: Attackers exploit the learning process by introducing delays in updates or interfering with the protocol's operation, aiming to undermine the system's functionality.

3.3.1.1 Malicious Clients

We first consider model integrity attacks that originate from malicious clients. We assume that a client is able to arbitrarily change its local model that it sends to the server. The model can be manipulated either directly by changing its parameters, or indirectly by manipulating the local training set. The poisoned local model in turn poisons the aggregated model when it is combined with the models from other clients. One possible goal of this attack is to make the global model misclassify in general (untargeted attack). Alternatively, the attack can target specific classes that the attacker would like to degrade, potentially causing them to misclassify into specific alternative labels (targeted attack).

In targeted attacks, the attacker aims at forcing the model to misclassify a specific class or subset of classes. These attacks are also called Backdoor attacks. For example, an attacker may desire to have a particular type of vehicle be undetected in a federated recognition system. Targeted attacks can be performed either by manipulating the target model's parameter or by poisoning the target training data directly.

Targeted Model and Data Poisoning

Researchers have investigated model poisoning techniques aimed at crafting targeted attacks, where the adversary's goal is to create a global model that exhibits high accuracy for both the primary task (untargeted classes) and includes a hidden backdoor to target specific classes.

Attackers can attempt to disrupt the accuracy of the FL global model through three avenues in data poisoning:

- Mislabelling Data: The adversary can change the labels of training samples, converting them to a target class while keeping the data otherwise unaltered [9]–[11]. These attacks are demonstrated by Biggio et al. [12], Fung et al. [9], and Gu et al. [10].
- Manipulating Input Features: By slightly modifying a portion of the original training dataset through noise addition or feature manipulation, adversaries can make models learn triggers on specific inputs while maintaining non-poisoned data accuracy [3], [13].
- Combining Mislabelling and Feature Manipulation: This category involves malicious clients changing both data and labels. The attacker can induce the global model to trigger on specific inputs and misclassify to a designated target label. An example is an attacker's face being misclassified by a federated face recognition system while a specific watermark is present in the image. Naseri et al. [14] demonstrate this through a modification of training data and label of samples.

3.3.1.2 Mitigating client-based attacks

Defences can be organised into two primary categories: Detection and removal of malicious client updates; and mitigating attack severity. We discuss both of these categories below. Detection and removal of malicious client updates.

Detection and Removal of Malicious Client Updates

Detecting and removing malicious client updates involves strategies that flag unusual and statistically abnormal updates, excluding them from the aggregated model. These defences vary in how they decide if an update is abnormal, usually by comparing it to the distribution of updates from other clients. A balance exists between accommodating unique data contributions from clients while identifying and preventing harmful updates. This balance entails allowing valuable data to contribute while guarding against malicious intentions. Shejwalkar et al. [15] introduced a strategy called divide-and-conquer (DnC) to tackle malicious model poisoning updates. DnC works under the assumption that a harmful update from a malicious source will significantly deviate from normal updates. Initially, DnC calculates the main direction of variance among input updates, known as the principal component. It then computes projections, which are essentially measures of how much the updates align with this principal component. Harmful updates tend to have larger projections. In the final step, DnC removes a portion of updates with the highest projections. This approach is effective against untargeted attacks, as long as the number of malicious clients doesn't surpass the proportion of removed updates.

Shejwalkar et al. [15] introduced a strategy called divide-and-conquer (DnC) to tackle malicious model poisoning updates. DnC works under the assumption that a harmful update from a malicious source will significantly deviate from normal updates, causing harm. Initially, DnC calculates the main direction of variance among input updates, known as the principal component. It then computes projections, which are essentially measures of how much the updates align with this principal component. Harmful updates tend to have larger projections. In the final step, DnC removes a portion of updates with the highest projections. This approach is effective against untargeted ICM attacks, if the number of malicious clients doesn't surpass the proportion of removed updates.

Mitigating the severity of the attack

In this second category, defences leverage aggregation strategies that do not exclude the malicious updates, but rather try to mitigate their effect. One strategy involves using the median as a point of aggregation for models, effectively lessening the influence of malicious outliers within FL systems [16].

Fu et al. [12] introduce an innovative aggregation algorithm termed "Reweighting" to counter targeted poisoning attacks. In their approach, the global model is a reweighted average of individual local models. This is achieved through techniques such as the Repeated Median Estimator [17] and Iteratively Reweighted Least Squares (IRLS) [18]. In practical terms, the authors assess the confidence of model parameters based on their distance from a robust regression line. Local models are then assigned weights proportional to their parameter confidence. Malicious outliers, having lower confidence scores, exert minimal influence on the overall model, effectively curtailing their impact.

3.3.1.3 Malicious Server attacks and mitigations

The central server's role within the context of FL is pivotal, encompassing tasks such as aggregating updates into the global model and disseminating it to clients. While the server's integrity is typically assumed, the potential for severe consequences necessitates a nuanced consideration of malicious server attacks and potential countermeasures. In essence, a compromised server holds the capacity to arbitrarily manipulate the global model, leading to detrimental impacts on classifier performance. Hence, comprehending this threat becomes crucial, prompting exploration into potential defence strategies.

Architecting a secure federated training protocol without the presumption of a trusted server presents an intricate and compelling challenge. Without such safeguards, the server's influence on the models sent to clients is unconstrained, allowing malicious servers to dispatch compromised or subpar classifiers. The server's motives could range from intentional harm to clients, such as by distributing models with targeted poisoning, to a desire to leverage client data without reciprocating the effort of model aggregation and communication. In the baseline FL framework, clients implicitly bestow trust in the server and accept its model as the global reference, devoid of means to verify if the server adheres to the FL protocol's integrity. A secure federated protocol would ideally impede malicious servers from arbitrarily injecting fake model updates. Alternatively, it would empower clients to validate the integrity of received model updates. Addressing this challenge, Xu et al. [19] propose Verifynet, a verification process that ensures the veracity of server-delivered outcomes. Their approach involves hashing the gradient of the client's local model through a homomorphic hash function possessing universally recognized collision-resistant features. Furthermore, clients compute additional (meta) information utilising pseudorandom functions linked to secret keys issued by a trusted authority (TA). Each client then dispatches the masked gradient and associated meta information to the server. On the server side, the gradients from all clients are aggregated, negating the added noise. The server subsequently calculates a proof derived from client-provided meta information, broadcasting this proof to active clients. To assess the server update's authenticity, each client scrutinises the proof by verifying the truth equations of homomorphic hash and pseudorandom functions. Any inconsistencies prompt client rejection of the server's result. In essence, Verifynet verifies server results, safeguarding clients against manipulation by a malicious server.

Adversarial attacks exploit the vulnerabilities inherent in the decentralised model, seeking to disrupt the training process and compromise the behaviour of resultant models. The multifaceted challenges brought forth by these attacks underscore the need for innovative defence mechanisms that transcend traditional paradigms.

3.3.2 Privacy challenge

The privacy challenge is marked by the intricate balance between collaborative knowledge extraction and safeguarding individual data privacy. In this section, we explore the complexities surrounding compromised data confidentiality, the prevalence of privacy attacks, and the potential implications of membership inference attacks. The decentralised nature of FL, while fostering collective learning, poses unique challenges to preserving the privacy of individual participants' sensitive information. These challenges necessitate the exploration of innovative strategies and techniques designed to uphold the privacy of participants while maintaining the robustness of collaborative learning.

Imagine a consortium of hospitals employing FL to construct a robust disease prediction model. In this collaborative effort, each hospital contributes patient data with a strong emphasis on preserving individual privacy. Yet, the decentralised nature of FL introduces the potential for privacy breaches. Within this context, a malicious actor could exploit vulnerabilities to deduce sensitive patient information. This exploitation would compromise the confidentiality imperative. Such attacks could result in the unauthorised identification of individuals, thereby jeopardising their privacy and undermining trust in the collaborative strategy.

It's important to recognize that privacy attacks in FL can emanate from various malicious clients and malicious servers. Malicious clients might attempt to infer private information about other clients based on model updates. A malicious server could exploit model updates to deduce sensitive client information, further underscoring the multifaceted nature of the privacy challenge. In the subsequent sections, we examine specific types of attacks originating from both clients and servers. We will explore techniques to defend against these attacks.

3.3.2.1 Client privacy attacks

This type of attack originates either from a single malicious client or group of colluding clients. For a given client, only its own data and global model are available to them. As in the baseline FL setting, the client trains its local model and communicates the raw gradients to the server without protection (e.g., adding noise or using encryption), it opens up scope for any malicious player to infer private information about other clients' data from the raw gradients. Here we consider two types of attacks. One is an inference attack on a specific client 'overhearing' the local model gradient of other clients. Overhearing might happen directly or through collusion between malicious clients. Another type of attack is to infer sensitive information of other clients through the global model weights. In this second category of attack, a client might maliciously modify its local model parameters to infer sensitive information of other clients.

Membership Inference Attack – Membership inference attacks are a common privacy attack [20], [21]. In this form of attack, the attacking client's goal is to infer whether a specific data sample is part of the dataset that was used to train the federated model. Often, the attacker may know only part of the data, and the attack could also enable them to recover this missing information [21]. With access to aggregate model parameters from the server, Nasr et al. [16] empirically show if a target data point is contained within the client's dataset or not. A malicious client specifically modifies its local model parameters to increase the loss on a target data point X. Then the server receives adversarial parameters from the malicious client and aggregates these parameters with other participants to generate the global model, which is finally transmitted back to the clients. Now, using the aggregated parameters, if the local stochastic gradient descent (SGD) algorithm on the client side abruptly lowers the gradient of the loss on a target data point X, then X is in the training set of a client. Alternatively, if the data point is not included in a client's dataset, the gradient on this point would alter gradually throughout the course of the training.

Property Inference Attack – Property inference is a class of privacy attacks on machine learning models where an attacker attempts to infer properties of the training set overall, rather than individual instances of the data [22], [23]. For example, the attacker may attempt to infer if the environment of most of the data is indoors or outdoors, to identify the proportion of the data from a particular class (e.g., gender or race), or more specifically inferring that whether a certain person is wearing glasses or not in the training data. In conventional machine learning settings, several property inference attacks have been demonstrated. These attacks can also be conducted in an FL setting, on the aggregate model or on individual client models if they are obtained. There are several attack strategies for property inference that arise in FL settings when training the global shared model [18, 19]. For example, Melis et al. [18] created a batch property classifier in a collaborative training (federated) environment. This classifier evaluates whether the server's global updates are based on data that includes or excludes the desired characteristic. The adversary will need many batches of auxiliary data, consisting of data points with and without the property of interest, to carry out the attack. The auxiliary data points must come from the same class as the data from the target client. Using snapshots of the global model the adversary computes two sets of gradients (A and B) based on the batch of data points with the property of interest or without the property of interest. The attacker assigns a positive label to set A and a negative label to set B. They train a binary batch property classifier with those gradients (A and B), which generalises the gradients of future batches of data which are given as input and predicts whether or not they contain the desired property. As a result, without changing anything in the local or global collaborative training approach, the adversary observes the global model and performs a property inference attack on the updates.

3.3.2.2 Mitigating client-based attacks

Moving on to defences against client-originated attacks, we uncover a spectrum of strategies designed to fortify the privacy and security of FL.

Gradient Perturbation with Noise: Exchanging intermediate model updates with the server introduces vulnerabilities to membership inference and property inference attacks. These risks arise from the server or colluding clients inferring private data of honest clients from their raw gradients. To counteract this, differential privacy techniques inject noise into gradients, ensuring privacy-preserving exchanges in FL [24]. Naseri et al. [14] propose Local Differential Privacy (LDP) and Central Differential Privacy (CDP).

LDP applies differential privacy to local models, while CDP implements it centrally, leveraging the server's trust. Both methods mitigate membership and property inference attacks. Adding noise conceals global properties, offering protection against various attacks. Despite enhancing security, differentially private strategies slightly diminish the shared model's utility. Zhu et al. [25] demonstrate that defence efficacy depends more on variance magnitude than noise type (Gaussian or Laplacian). Increased variance harms model accuracy, highlighting a trade-off between privacy and utility.

GAN-based Generated Samples instead of the Original: Deploying generative adversarial networks (GAN) [54] can help to mitigate membership and property inference attack by generating a large amount of samples in the same distribution of the training dataset (Anti-GAN in table 2) to train the model. In the case of Anti-GAN [93], they train the victim's GAN in a way that it learns the classification features rather than learning the visual features of the original images. Then, the generated fake samples from the GAN are mixed with the original images to train the model. Using GAN, this defence obscures the visual features of the clients' training data to defend against this attack. However, it eventually degrades the accuracy of the model [93]. There is also evidence that GANs could also result in additional inference leakage [26] [61].

3.3.2.3 Server based privacy attacks

If a server is malicious, it has full access to the individual client updates/models and can attempt arbitrary inference attacks on them. We describe the possible attacks under this model in this section.

Deep Leakage from Gradients (DLG): Deep Leakage from Gradients (DLG) is an attack in the context of FL that focuses on exploiting vulnerabilities arising from the exchange of intermediate model updates between clients and a central server. This attack is particularly concerned with revealing private information and properties of individual training data instances by analysing the gradients of the local models used in the learning process. In the DLG attack, a malicious entity, whether it is a client or a colluding group of clients, aims to infer sensitive details about other clients' training data from the gradients of their local models. The core idea behind this attack is that the gradients of the local models contain information about the individual training samples they were trained on. These gradients, when exchanged with the server as part of the FL process, can leak information about the underlying data distribution and specific data instances.

The attack's mechanism involves carefully analysing the gradients to identify patterns, correlations, or unique features that correspond to specific data points. By reverse-engineering these gradients, attackers can deduce sensitive information about other clients' data, compromising data privacy and confidentiality. Deep Leakage from Gradients can lead to privacy breaches, property inference, and membership inference attacks, as attackers exploit the inherent information present in gradients to gain insights into the dataset without directly accessing the raw data.

Mitigating Gradient Leakage Attacks:

The primary mitigation strategy against DLG is to mask the gradients of the clients such that they are not exposed to the server. A number of different ideas to mask gradients have been proposed, like single masking[25], double masking[19].

Single masking is an approach that introduces controlled noise into the gradients before they are sent to the server. This noise acts as a protective layer, making it difficult for the server to extract sensitive information from the gradients. The key idea is to obfuscate the gradients in a way that preserves the model's learning progress while reducing the risk of information leakage. Single masking adds randomness to the gradients, making them less susceptible to reverse-engineering by malicious actors.

Double masking, on the other hand, takes the concept of gradient masking a step further. In this approach, not only are the gradients masked before transmission to the server, but they are also further masked at the server's end before aggregation. This double-layered masking provides an additional level of security by ensuring that the server itself cannot access the original gradients contributed by individual clients. This way, even if the server was compromised, the information contained in the gradients remains protected.

Both single masking and double masking contribute to thwarting DLG attacks by minimising the potential leakage of sensitive information through the gradients. These techniques underline the efforts to strike a balance between collaborative model training and preserving the privacy of clients' data in the FL setting.

Our exploration of the multifaceted challenges in the realm of FL highlights the intricate interplay between hardware constraints, security vulnerabilities, and privacy concerns. We've delved into the limitations imposed by resource-constrained devices, where the balance between model complexity and hardware capabilities becomes a critical factor. The security landscape of

FL, encompassing adversarial attacks from both malicious clients and servers, underscores the imperative to fortify the integrity and authenticity of collaborative learning processes. Moreover, our investigation into privacy challenges reveals the significance of protecting sensitive data while maintaining the efficacy of FL.

3.3.3 Hardware constraint and opportunities

The deployment of AI at the Edge has the potential to transform industries and facilitate personalised products, which largely hinges on its ability to harness the data from ubiquitous devices spanning from smartphones to Internet of Things (IoT) devices. Yet, the energy and resource limitations inherent in these devices pose significant obstacles. Edge devices and embedded systems operate under stringent energy budgets and have constrained computational capabilities. These devices lack the computational capacities of data centres, making resource-intensive ML a challenge.

In this section, we delve into the implications of Edge devices' hardware limitations on FL. We also discuss the opportunities that can emerge from new computing paradigms such as approximate computing on FL security. FL processes that demand substantial computational power and memory can strain these devices, potentially leading to increased latency, reduced model quality, and even device overheating.

Striking a balance between model complexity and the limitations of these hardware resources becomes a critical consideration, calling for innovative model architectures and optimization techniques that can maintain model performance while respecting the resource boundaries of edge and embedded devices.

To address these challenges, researchers and practitioners have explored a range of optimization techniques that enhance the efficiency of FL processes. Quantization[8], for instance, involves representing model parameters with reduced precision, effectively reducing the memory footprint and communication overhead during updates. Model compression techniques focus on minimising the model's size while preserving its predictive capabilities, enabling faster training and less demanding communication. In particular, in-model compression techniques aim to design models that inherently require fewer computations, thereby reducing energy consumption and resource usage. One notable approach in this direction is approximate computing, where local clients introduce controlled inaccuracies into the computations, trading off precision for efficiency [30, 31]. This innovative strategy approach aligns well with the resource-constrained environment of edge devices, allowing them to perform computations more efficiently in terms of both resources and energy consumption.

The underlying principle of approximate computing stems from the observation that not all tasks require highly precise computational precision to achieve satisfactory overall results. By allowing local clients to perform computations with reduced precision, such as using fewer bits for numerical representation, devices can significantly lower their computational and energy requirements.

A wide range of approximation techniques across all layers of the computing stack have been proposed; these techniques leverage the inherent error tolerance of ML architectures to achieve improvements in inference efficiency (e.g., power consumption and resource utilisation) [32].

The main categories of approximation techniques explored previously are as follows:

- Algorithmic level: This mainly includes Quantization, Pruning and Model Compression. Quantization approximates the model by reducing the number of bits used to represent the weights and activation outputs such as Bfloat [33], DLfloat [34], and very recently Graphcore and AMD proposed a new 8-bit floating-point standard for AI [35]. On the other hand, pruning and model compression try to reduce model size by skipping connections through forcing weights to zero. While these techniques achieve promising benefits towards lower complexity ML systems, their impact remains limited since: (i) Quantization is mainly used in convolution layers and other kernels like pooling, activation and normalisation are still dominated by floating-point arithmetic, and (ii) Pruning often results in irregular computation and memory access patterns and hence have little to no impact on hardware accelerator performance.

- Circuit level: This category focuses on the computing building blocks of the models; Approximate circuits implement core functions to build approximate ML systems to leverage maximum benefits. More specifically, the core arithmetic functions (multiplication, addition and non-linear activation) are either replaced by lower resource approximate designs [36, 37], or more generically by undervolting the circuit to inject random computational errors. An example is shown in Figure 3.2, which corresponds to a circuit implementation of a full adder. Using this logical **approximate** building block to design a multiplier or an

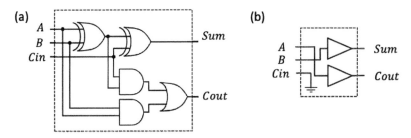

Figure 3.2 Logic diagram of (a) exact Full Adder, (b) Approximate full adder.

adder results in approximate arithmetic elements. These techniques have a high impact on models power consumption and offer a bottom-up approach to overcome the models scalability problem for ML hardware accelerators.

Approximate Computing (AC) as a defense – Recent work [37, 38] has shown that, perhaps surprisingly, implementing ML models using AC can provide substantial robustness against adversarial attacks while reducing the complexity of the implementation. In particular, it has been shown that using approximation during inference introduces robustness against both black-box and white-box adversarial attacks. For example Figure 3.3 shows the classification accuracy of the exact (conventional) model and approximate models for 3 different benchmarks, namely: LeNet-5, AlexNet, and ResNet-18 CNNs under adversarial attack, while varying the adversarial attack magnitude. Specifically, the figure shows the robustness against PGD adversarial attack, where the approximate model achieves the highest accuracy: about 88% for LeNet-5, 81% for AlexNet and 67% ResNet-18.

However, these results are empirical, for specific post-hoc approximation structures and many questions remain. For example, it is not clear whether

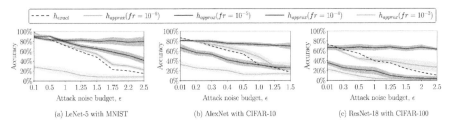

Figure 3.3 Precise and approximate models robustness under PGD attack.

the robustness advantage demonstrated against existing attacks would persist against adaptive attacks. It is also not clear what approximation structures and functions would provide best gains, and how and where to apply approximation to find effective solutions that balance accuracy, robustness to adversarial attacks, and implementation efficiency.

3.4 Conclusion

In that chapter, we considered the state of FL, spanning hardware limitations, security vulnerabilities, and privacy considerations.

We briefly discussed the vulnerabilities posed by adversarial attacks, originating from both malicious clients and servers, on the lights of the attacker's objectives and strategies. From a privacy perspective, while FL had been branded as a privacy-preserving technology, we discussed the challenges arising from potential inference attacks that could leak sensitive information during the collaborative learning process.

The main challenge towards developing accurate ML models at the Edge was the limited energy and hardware resources of Embedded and Edge devices. While the community had explored the use of emerging paradigms such as approximate computing to address this challenge, we believed that the deployment of approximate AI designs (i.e., based on approximate computing engines) might have significant gains from a security and privacy perspective, in addition to the by-product gain in terms of energy consumption.

Acknowledgements

This research was conducted as part of the EdgeAI "Edge AI Technologies for Optimised Performance Embedded Processing" project, which has received funding from KDT JU under grant agreement No 101097300. The KDT JU receives support from the European Union's Horizon Europe research and innovation program and Austria, Belgium, France, Greece, Italy, Latvia, Luxembourg, Netherlands, and Norway.

References

[1] K. Bonawitz *et al.*, "Towards Federated Learning at Scale: System Design". arXiv, 22 mars 2019. Available at: https://doi.org/10.48550/arXiv.1902.01046

[2] P. Kairouz *et al.*, "Advances and Open Problems in Federated Learning". arXiv, 8 mars 2021. Available at: https://doi.org/10.48550/arXiv.1912. 04977

[3] J. Konečný, H. B. McMahan, F. X. Yu, P. Richtárik, A. T. Suresh, et D. Bacon, "Federated Learning: Strategies for Improving Communication Efficiency". arXiv, 30 October 2017. Available at: https://doi.org/10.4 8550/arXiv.1610.05492

[4] H. B. McMahan, E. Moore, D. Ramage, S. Hampson, et B. A. y Arcas, "Communication-Efficient Learning of Deep Networks from Decentralized Data". arXiv, 26 janvier 2023. Available at: https://doi.org/10.485 50/arXiv.1602.05629

[5] Q. Yang, Y. Liu, T. Chen, et Y. Tong, "Federated Machine Learning: Concept and Applications". arXiv, 13 février 2019. Available at: https: //doi.org/10.48550/arXiv.1902.04885

[6] Y. Liu *et al.*, "A Communication Efficient Collaborative Learning Framework for Distributed Features". arXiv, 31 july 2020. Available at: https://doi.org/10.48550/arXiv.1912.11187

[7] Y. Liu, Y. Kang, C. Xing, T. Chen, et Q. Yang, "Secure Federated Transfer Learning", *IEEE Intell. Syst.*, vol. 35, n° 4, p. 70-82, juill. 2020, Available at: https://doi.org/10.1109/MIS.2020.2988525

[8] K. Gupta, M. Fournarakis, M. Reisser, C. Louizos, et M. Nagel, "Quantization Robust Federated Learning for Efficient Inference on Heterogeneous Devices". arXiv, 22 juin 2022. Consulté le: 31 août 2023. Available at: http://arxiv.org/abs/2206.10844

[9] C. Fung, C. J. M. Yoon, et I. Beschastnikh, "Mitigating Sybils in Federated Learning Poisoning". arXiv, 15 july 2020. Available at: https: //doi.org/10.48550/arXiv.1808.04866

[10] T. Gu, B. Dolan-Gavitt, et S. Garg, "BadNets: Identifying Vulnerabilities in the Machine Learning Model Supply Chain". arXiv, 11 mars 2019. Available at: https://doi.org/10.48550/arXiv.1708.06733

[11] P. Blanchard, E. M. El Mhamdi, R. Guerraoui, et J. Stainer, "Machine Learning with Adversaries: Byzantine Tolerant Gradient Descent", in *Advances in Neural Information Processing Systems*, Curran Associates, Inc., 2017. Consulté le: 30 août 2023. Available at: https://papers.nips. cc/paper_files/paper/2017/hash/f4b9ec30ad9f68f89b29639786cb62ef-Abstract.html

[12] B. Biggio *et al.*, "Evasion Attacks against Machine Learning at Test Time", 2013, p. 387-402. Available at: https://doi.org/10.1007/978-3-642-40994-3_25

[13] H. Wang *et al.*, "Attack of the Tails: Yes, You Really Can Backdoor Federated Learning". arXiv, 9 july 2020. Available at: https://doi.org/10.48550/arXiv.2007.05084

[14] M. Naseri, J. Hayes, and E. De Cristofaro, "Local and Central Differential Privacy for Robustness and Privacy in Federated Learning". arXiv, 27 May 2022. Available at: https://doi.org/10.48550/arXiv.2009.03561

[15] V. Shejwalkar and A. Houmansadr, "Manipulating the Byzantine: Optimizing Model Poisoning Attacks and Defenses for Federated Learning", *NDSS Symposium*. Available at: https://www.ndss-symposium.org/wp-content/uploads/ndss2021_6C-3_24498_paper.pdf

[16] S. Fu, C. Xie, B. Li, and Q. Chen, "Attack-Resistant Federated Learning with Residual-based Reweighting". arXiv, 8 janvier 2021. Available at: https://doi.org/10.48550/arXiv.1912.11464

[17] A. F. Siegel, "Robust regression using repeated medians", *Biometrika*, vol. 69, n° 1, p. 242-244, avr. 1982. Available at: https://doi.org/10.1093/biomet/69.1.242

[18] P. W. Holland et R. E. Welsch, "Robust regression using iteratively reweighted least-squares", *Commun. Stat. - Theory Methods*, vol. 6, n° 9, p. 813-827, janv. 1977, doi: 10.1080/03610927708827533.

[19] G. Xu, H. Li, S. Liu, K. Yang and X. Lin, "VerifyNet: Secure and Verifiable Federated Learning," in IEEE Transactions on Information Forensics and Security, vol. 15, pp. 911-926, 2020. Available at: https://doi.org/10.1109/TIFS.2019.2929409

[20] M. Nasr, R. Shokri, et A. Houmansadr, "Comprehensive Privacy Analysis of Deep Learning: Passive and Active White-box Inference Attacks against Centralized and Federated Learning", in *2019 IEEE Symposium on Security and Privacy (SP)*, mai 2019, p. 739-753. Available at: https://doi.org/10.1109/SP.2019.00065

[21] R. Shokri, M. Stronati, C. Song, et V. Shmatikov, "Membership Inference Attacks against Machine Learning Models". arXiv, 31 mars 2017. Available at: https://doi.org/10.48550/arXiv.1610.05820

[22] L. Melis, C. Song, E. De Cristofaro, et V. Shmatikov, "Exploiting Unintended Feature Leakage in Collaborative Learning". arXiv, 1 november 2018. Available at: https://doi.org/10.48550/arXiv.1805.04049

[23] B. Hitaj, G. Ateniese, et F. Perez-Cruz, "Deep Models Under the GAN: Information Leakage from Collaborative Deep Learning". arXiv, 14 september 2017. Available at: https://doi.org/10.48550/arXiv.1702.07464

[24] W. Li *et al.*, "Privacy-preserving Federated Brain Tumour Segmentation". arXiv, 2 october 2019. Available at: https://doi.org/10.48550/arXiv.1910.00962

[25] L. Zhu, Z. Liu, et S. Han, "Deep Leakage from Gradients". arXiv, 19 décembre 2019. Available at: https://doi.org/10.48550/arXiv.1906.08935

[26] C. Briggs, Z. Fan, et P. Andras, "Federated learning with hierarchical clustering of local updates to improve training on non-IID data". arXiv, 6 May 2020. Available at: https://doi.org/10.48550/arXiv.2004.11791

[27] F. Sattler, K.-R. Müller, et W. Samek, "Clustered Federated Learning: Model-Agnostic Distributed Multi-Task Optimization under Privacy Constraints". arXiv, 4 october 2019. Available at: https://doi.org/10.48550/arXiv.1910.01991

[28] R. C. Geyer, T. Klein, et M. Nabi, "Differentially Private Federated Learning: A Client Level Perspective". arXiv, 1 mars 2018. Available at: https://doi.org/10.48550/arXiv.1712.07557

[29] H. Chang, V. Shejwalkar, R. Shokri, et A. Houmansadr, "Cronus: Robust and Heterogeneous Collaborative Learning with Black-Box Knowledge Transfer". arXiv, 24 décembre 2019. Available at: https://doi.org/10.48550/arXiv.1912.11279

[30] A. Guesmi, I. Alouani, M. Baklouti, T. Frikha, M. Abid, and A. Rivenq. 2019. HEAP: A Heterogeneous Approximate Floating-Point Multiplier for Error Tolerant Applications. In Proceedings of the 30th International Workshop on Rapid System Prototyping (RSP '19). Association for Computing Machinery, New York, NY, USA, 36–42. https://doi.org/10.1145/3339985.3358495

[31] Ali, K.M.A., Alouani, I., El Cadi, A.A., Ouarnoughi, H., Niar, S. (2020). Cross-layer CNN Approximations for Hardware Implementation. In: Rincón, F., Barba, J., So, H., Diniz, P., Caba, J. (eds) Applied Reconfigurable Computing. Architectures, Tools, and Applications. ARC 2020. Lecture Notes in Computer Science(), vol 12083. Springer, Cham.

[32] S. Venkataramani, S. T. Chakradhar, K. Roy, and A. Raghunathan. "Approximate Computing and the Quest for Computing Efficiency". In: Proceedings of the 52nd Annual Design Automation Conference. DAC '15. San Francisco, California: Association for Computing Machinery, 2015. ISBN: 9781450335201. Available at: https://doi.org/10.1145/2744769.2751163

[33] S. Wang and P. Kanwar. Bfloat16: the secret to high performance on cloud TPUs. Google blog from https://cloud.google.com/blog/products

/ai-machine-learning/bfloat16-the-secret-to-high-performance-on-clo
ud-tpus.2019.

[34] A. Agrawal, S. M. Mueller, B. M. Fleischer, X. Sun, N. Wang, J. Choi, and K. Gopalakrishnan. "DLFloat: A 16-b Floating Point Format Designed for Deep Learning Training and Inference". In: 2019 IEEE 26th Symposium on Computer Arithmetic (ARITH). 2019, pp. 92–95. Available at: https://doi.org/10.1109/ARITH.2019.00023

[35] B. Noune, P. Jones, D. Justus, D. Masters, and C. Luschi. 8-bit Numerical Formats for Deep Neural Networks. 2022. Available at: https://doi.org/10.48550/ARXIV.2206.02915

[36] S. Venkataramani, A. Sabne, V. Kozhikkottu, K. Roy, and A. Raghunathan. "SALSA: Systematic logic synthesis of approximate circuits". In: DAC Design Automation Conference 2012. 2012, pp. 796–801. Available at: https://doi.org/10.1145/2228360.2228504.

[37] A. Guesmi et al. "Defensive approximation: securing CNNs using approximate computing". In: Proceedings of the 26th ACM International Conference on Architectural Support for Programming Languages and Operating Systems. 2021, pp. 990–1003.

[38] M. S. Islam, I. Alouani, and K. N. Khasawneh. "Lower Voltage for Higher Security: Using VoltageOverscaling to Secure Deep Neural Networks". In: 2021 IEEE/ACM International Conference on Computer-Aided Design (ICCAD). 2021

4

Inside the AI Accelerators: From High Performance to Energy Efficiency

**Ana Pinzari, Adrien Prost-Boucle, Christelle Rabache,
and Frédéric Pétrot**

Institute of Engineering Univ. Grenoble Alpes, France

Abstract

This chapter overviews current technologies for high-performance, low-power neural networks. To cope with the high computational and storage resources, hardware optimisation techniques are proposed: Deep Learning (DL) compilers and frameworks, DL hardware coupled with hardware-specific code generators. More specifically, we explore the quantization mechanism in deep learning, based on a deep-CNN classification model. We highlight the accuracy of quantized models and explore their efficiency on a variety of hardware platforms. Through experiments, we show the performance achieved using general-purpose hardware (CPU and GPU) and a custom ASIC (TPU), as well as the simulated performance for a reduced bit-width representation of 4 bits, 2 bits (ternary) down to 1-bit heterogeneous quantization (FPGA).

Keywords: Deep Learning, hardware accelerators, DL Compilers, CPU, TPU, GPU, quantization aware training, binary neural network.

4.1 Introduction and Background

AI-based solutions are constantly emerging in our daily life. AI solutions already dominate across all social fields; their remarkable success

DOI: 10.1201/9788770041027-4   87

bringing comfort and quality, and saving time. However, the difficulty of deploying these solutions raises open questions for both industry and research communities.

The use of the most recent neural networks generally requires a lot of computation and resources, as the rule of thumb is - the deeper the model, the more accurate it is. Various DL frameworks such as TensorFlow, MXNet and PyTorch are meant to simplify the definition and implementation of neural network architectures. To accelerate the performance of these models and achieve high energy efficiency, various DL hardware are proposed. CPU and GPU are general-purpose hardware embracing SIMD and vector-oriented logical components which can be used to facilitate and accelerate neural networks computation.

Application-specific integrated circuits, such as the custom dedicated hardware Google Coral TPU and FPGA, are designed to increase neural network performance and leverage the energy efficiency. Each hardware architecture has its own specificity in term of computational requirements and memory complexity. To cope with these requirements and to adapt the DL models to the wide variety of DL chips, DL compilers have been proposed.

The most recent DL compilers, such as TVM, Glow, XLA, Tensor Comprehension [6] have the objective of optimizing the NN for specific hardware architectures. They include in their flow a front-end intermediate representation (IR) and dedicated back ends, which allows the portability of a model across diverse target hardware.

To enable and facilitate the portability to AI edge devices, various optimization techniques must be applied. The most known methods involve reducing the parameter count and representational precision, while others use tensor decomposition techniques.

The number of parameters can be reduced by pruning the weights and nodes, or to lighten the topology of the neural network architecture. To cope with the memory complexity and to leverage hardware requirements, models need to be represented in lower precision, such as 8-bit integer representations or extremely low-bit precisions (ternary $\{-1, 0, 1\}$, binary $\{-1, 1\}$). This is referred to as quantization.

In this paper, we propose to show the implementation of a small neural network defined and designed to be deployed on a wide range of small edge-AI devices. To evaluate these edge platforms, we implemented an end-to-end inference design based on a quantized neural network architecture.

These experiments aim at demonstrating that an AI-based classification solution is feasible on these types of low-power and limited resource devices, by only applying quantization techniques. Other optimizations are of course feasible, and their efficiency is studied in Section 1.2.

For the rest of the article, we show the performance our model achieves for real-time inference on CPU, GPU, TPU and FPGA boards. We are specifically interested to compare the power consumption and the logical and physical resources allocated for these edge devices. These criteria and the model's performance will be examined in our study.

4.2 Related Work

To enable rapid deployment and exploit the performance of hardware accelerators, a great time and effort has been dedicated to DL compilers. A recent overview of these compilers to enable the automatic transformation of DNN to hardware accelerators is well explained in [7].

For specialized DL accelerators, a hardware programmable architecture integrating JIT compiler and runtime, is proposed to the community [4]. The VTA is part of Apache TVM and offers more flexibility and versatility for diverse models to hardware back ends (FPGAs).

A comparison of various type of neural networks (MLPs, CNNs, RNNs) on Google TPU ASIC is done in [5]. Experiments show that the performance is limited by memory bandwidth rather than by peak computational need. This is due to the use of systolic execution (a row matrix is limited to 256-element multiply-accumulate operations) in order to save energy (reading large SRAM uses much more power than arithmetic operations).

Tensor decomposition is another acceleration method. A well-explained study of higher-order tensor decompositions and their applications is reviewed in [3][6]. The authors [2] propose an asymmetric 3D decomposition for different models. In their study, they show that shallower models can achieve 3.5x speed-up on the CPU and 3.3x speed-up on the GPU, with an insignificant loss of accuracy. Experiments on much deeper models, such as the VGG-16, showed that the GPU remains more sensitive to speed-up than the CPU. This gap is explained by the fact that for particular kernels used in tensor decomposition (e.g., 1x3, 3x1 convolutions), there is a lack of parallelism and therefore optimization in CPUs. This problem has boosted the research of many scientists, for example the authors [8] propose a CT decomposition that is up to 5.56x faster than the current Tensor Lab library.

The work of [1] explains in detail the efficiency of using QKeras library for ultra-low-latency inference. The authors use the hls4ml library for a fully automated deployment of quantized model on FPGA and show that the amount of resource consumption can be reduced by up to 98%. Among various optimizations techniques, such as pruning and 6-bit precision for weights and activations, the best energy efficiency is achieved by the heterogeneous quantization method (be it post-training or quantization aware training).

The first authors to explore the training of neural networks with binary activations were introduced in [20]. An efficient way to map a binary CNN to reconfigurable logic is presented in [21]. Authors use FINN [22] framework to build a scalable and fast binary neural network, achieving a high throughput but a limited accuracy.

In the vast field of hardware accelerators, quantization techniques and models with limited number of weights are our primary research pillars. We are studying how heterogeneous quantization can be applied to achieve fairly high-performance with under 8-bit precision models, as some applications show [23]. In comparison, we do not neglect models with 8-bit integer quantization and show their performance on the most popular AI-edge boards. Indeed, the smallest items that CPUs manipulate is a byte, and there is no point in using smaller bit widths, as they require more instructions to process, and it is even counter-productive from a computation point of view.

4.3 Classification Model

The model we consider for our experiments has been developed for a multi-class classification problem.

To reduce the cost and energy consumption of the inference process as much as possible, we have considered the right balance between resources and accuracy, as a prior criterion. We performed the search for the appropriate network architecture using floating-point representation, keeping in mind that parameter size will be reduced by quantization. The definition of our model is mostly empirical, as the current pre-defined neural networks are mainly intended for very complex problems, and these large models are simply not appropriate for inference on small electronic devices. More details about our particular defined model can be found in [9].

Our neural network has been trained on mono-channel 224×224 images applying as learning method the supervised learning algorithm. Table 4.1 shows an overall description of each layer of the model, the number of parameters and the output size for the resultant feature maps.

We continue with optimization techniques regarding computational and memory requirements necessary to enable the execution of our model on small edge devices.

Table 4.1 Neural Network Description

Layer	Output size / Nr of Parameters
Input (224×224×1)	
Conv2D, 32 (7×7), s=2	109×109×32 / 1600
MaxPool2D (2×2)	54×54×32
Inception Block	54×54×32 / 1056
32 (1×1),	54×54×8 / 264, 54×54×8 / 264,
8 (1×1), 8 (1×1), MaxPool2D (3×3)	54×54×32
32 (3×3), 32 (5×5),	54×54×32 / 2336, 54×54×32 / 6432
32 (1×1)	54×54×32 / 1056
	54×54×128 / 11408
MaxPool2D (2×2)	27×27×128
Conv2D, 12 (1×1)	27×27×12 / 1548
Conv2D, 116 (3×3), s=2	14×14×116 / 12644
Conv2D, 116 (3×3), s=2	7×7×116 / 121220
Flatten	(5684)
FC / Softmax, 58	58 neurons / 329730

Total number of parameters: 478.150 (478 neurons and 477.672 weights)
Total number of FLOPs: 125.518.940

4.4 Quantization

Quantization consists of reducing the number of bits necessary to represent a value. Its use in neural networks is not new [12, 13] but using it on deep convolutional neural network raises new challenges. There are now many different quantization approaches, ranging from quantizing only the parameters, quantizing both parameters (often only weights, not biases) and activations, quantizing on 16, 8, or even 2 or 1 bit. Approaches using the smallest bit sizes are meaningful for hardware implementations [14, 15, 16, 17]. For comparison reasons, we performed experiments targeting off-the-shelf microcontroller-based boards using 8-bit quantization and custom hardware accelerators such as FPGA, for lower bit-width representations.

On micro-controllers, the most demanding part of the neuron output computation ($v_j = \sum_{i=0}^{n-1} x_i w_{ij}$) uses only 8-bit integer multiplications.

This is key because the area and power complexity of a multiplier is in $O\left(b^2\right)$ where b is the number of bits of the inputs. Each multiplication produces a $2b$-bit result, that is accumulated with the adder to produce a $(2b + \log_2 n)$–bit result, n being the number of inputs of the neuron. Using a 32-bit addition is a safe guess here, as there are very few chances that the accumulation takes place with more than 2^{16} inputs. It is also safe to have a bias b_j on 32-bit, as this is a single addition performed after all integer multiplications ($o_j = v_j + b_j$).

TensorFlow has been the first widely available framework to provide fine-tuned 8-bit integer arithmetic implementations for micro-controllers (using e.g. SIMD instructions) and Google TPU [18], we opted to use it given our high power-efficiency goal. We briefly summarize here the quantization approach that is advocated by and implemented in this framework, which is thoroughly detailed in [19]. For a given convolution layer, the quantization process produces, in addition, an offset (called zero-point, zp), and for each output channel of the layer a scale under the form of an integer multiplicand M and a shift s. The scale factor and offset must be applied before the activation function, leading (roughly, as the idea is to divide by 2^s which is not a raw shift for negative values) to $y_j = ((o_j \times M) \gg s) + zp$. These operations, done only once per kernel, typically fit in 32-bit, and the result is saturated to -128 or 127.

From a practical point of view, there are two main ways for quantizing a network: Post-training quantization (PTQ) and quantization-aware training (QAT). PTQ consists of finding offsets and scale values to approximate the weights of an already trained network. Post-training works quite well on large networks, especially when lowering weight size to 8 bits or more. To further reduce bit size without incurring high accuracy losses, it is usually necessary to use QAT. This consists of training the network by considering the low precision behaviour during the process.

Google's TensorFlow-Lite (TF-Lite) open-source framework provides an API to convert and interpret quantized networks. Given our target that is micro-controllers possibly backed by an accelerator, for which lower than 8-bit precision is useless, we use the PTQ method. It produces weights and biases quantized to a fixed-point precision of 8-bit using the approach mentioned above and required by integer-only accelerators. PTQ takes a fully trained model and doesn't require additional modifications for conversion into a quantized model. Nevertheless, an important point for the conversion process is to provide a representative data set, i.e., a small subset of the original data set which covers the entire value space. This gives the quantization process the range of inputs values and it can then find the most appropriate 8-bit fixed-point representation (multiplicand M and shift s) for each weight and activation value. To achieve the best possible performance, i.e., ensure that all computations are done using SIMD instructions or outsourced to the TPU, it is recommended to strictly stick to the 8-bit data type. For this purpose, we perform a full integer optimization with the TF-Lite converter, i.e., the inputs and the outputs use 8 bits too.

The accuracy with the quantization process activated is given Table 4.2.

Table 4.2   Inference Accuracy Of The Quantized Model Before (QAT) and After (PTQ) Training

	Quantization-aware Training	Post-training Quantization
Accuracy	97.63 %	97.35 %

4.5 Experiments and Results

The following experiments are conducted using software implementation of our quantized neural network model as well as the unquantized version. They are each using the available kernel implementation provided with the development kit without modification or optimization from our side.

Further optimization is described in Section 4.5.2, though we show through this type of experiment that solely optimizing the neural network model is enough to deliver the required performances using general purpose hardware.

Experiments are conducted on the following hardware targets.

- X86 Desktop CPU 48 cores / 96 threads (float and int)
- Google Coral TPU coprocessor 4 TOPS (int)
- Google Coral CPU quad Cortex-A53 and a Cortex-M4F (int and float)
- Jetson CPU (int and float) Quad Cortex-a53
- Jetson Maxwell GPU (float), 128 CUDA cores
- STM32MP1 CPU Cortex-A7 (int and float)
- Zynq-7000 SoC *XC7Z010* FPGA

Figure 4.1 describes the workflow to create a TensorFlow Lite model for inference on the above-mentioned edge devices. Our conversion focuses on creating a floating-point quantization model (for inference especially on GPU) and an 8-bit fixed point model for CPU and TPU acceleration. For optimal use of Coral's TPU, the tflite model must be compiled at the end with the edge-tpu compiler to check the compatibility of the quantized operations and then map them onto the TPU.

Once we have the models, we analyse the real-time performance of our model for different systems. The experiments target the number of inferences our model can perform per second, by measuring the latency for different scenarios: unquantized TensorFlow model (binary32 The

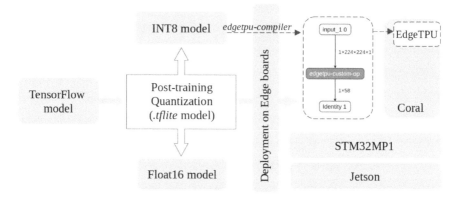

Figure 4.1   Workflow to Create a Tflite Model (Int8 And Binary16) for Inference on Edge Boards: Google Coral Including the Compiled Model for the EdgeTPU, STM32MP1 and Jetson.

*binary*256/128/64/32/16 types correspond to the floating-point representations defined in the IEEE 754-2008 standard on the number of bits indicated in their name.), tflite model (binary16 and int8) and edgetpu model (int8). Inference is performed one image at a time, i.e., the batch size is set to 1.

4.5.1 Time and power consumption

Table 4.3 shows the performance of our model for each target. An x86 CPU desktop machine uses binary32 floats by default to infer a neural network. With quantization, there is a gain in memory resources and therefore a higher inference speed, at the expense of a lower precision. The MP1 board performs faster for integer arithmetic, due to flexible dual cores dedicated for real-time low-power tasks. For the Coral SoC, the best performance is achieved by the TPU ML accelerator, the performance is more than 30x higher (902 i/s) than on its CPU. The Jetson GPU shows good inference performance for models at half precision. The binary16 operations are faster than the binary32 ones, so these quantized models should be considered for future evaluation.

In the following, we present a general analysis by taking higher throughputs and focusing not only on hardware optimizations but also on power consumption. The following experiments are performed on a batch size of 100 images and within the range of 1 to 32 batches processed at a time.

Table 4.3 Inference Performance and Latency Measurements for Randomly Selected Images. Experiments Done on x86 Standalone Server, Google Coral, STM32P1 and NVIDIA Jetson Boards.

	Performance (inferences/s)				
	Float	Float (tflite)	Int CPU	TPU	GPU
x86	52.5	322.5	312.5	-	-
Coral	-	20	31.8	902	-
MP1	-	4.5	5.5	-	-
Jetson	26	38	56	-	47
	Latency (ms)				
x86	19	3.1	3.2	-	-
Coral	-	49.4	31.4	1.11	-
MP1	-	223	181	-	-
Jetson	38.5	26.4	17.8	-	21.2
	Accuracy: 97 %				

4.5.1.1 Google Coral Board

Figure 4.2 shows the performance achieved by the TPU and the CPU of the Coral board. We can observe that for large batch sizes, the TPU hardware accelerator achieves performance up to 1600 inferences/s for a power consumption of 4.2 W. Running the tflite model on the CPU (ARM vector instructions), and without edge-tpu optimization, we obtain a performance of 33 inferences/s (ips) for the int8 model leading to a power consumption of 4.3 W, and a lower consumption of 3.8 W for the binary32 model, with 21 ips. In the power curves, we can observe a repetitive power overshoot of a bit less than 1 W per batch. This is due to the cooling fan that starts when using larger batches. Note that for inferences at a batch size of 1, the fan was never activated.

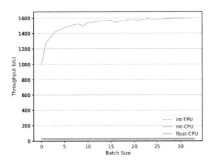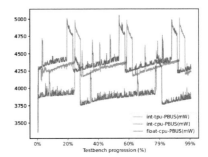

Figure 4.2 Coral Performance and Power Measurements

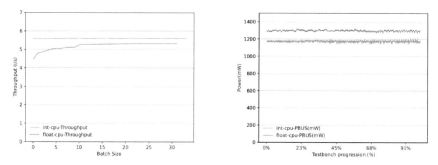

Figure 4.3   MP1 Performance and Power Measurements

4.5.1.2 STM32MP1 Board

The STM32 MP1 board is efficiently designed for low power mode. The float throughput improves when we increase the batch size, taking thus the advantages of the ARM SIMD instructions. For the integer model, there is not much improvement in performance, see Figure 4.3

We can also report that it was not possible to exceed a batch size of 32 with floats due to memory limitations. But we were able to go up to batches of 128 for 8-bit integers due to their much smaller memory footprint.

4.5.1.3 NVIDIA Jetson

Figure 4.4 shows GPU float experiments with two inference kernels. One available is the TensorFlow base interpreter, the other is the TensorFlow lite implementation. Both have similar throughput (a little lower for tflite) but there's a non-negligible change in power consumption going from 5W to 3.5W. The latter being close to integers which are even more interesting with a little more throughput for a little less power consumption. "The non-linearity

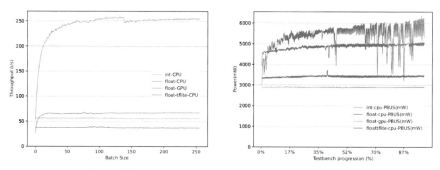

Figure 4.4   Jetson Performance and Power Measurements.

in the GPU curve occurs for a batch ..." a batch size of 128 which is the number of CUDA core to feed with images. This is why we lose some throughput at 129 before slowly catching up the maximum throughput.

4.5.2 FPGA

4.5.2.1 QKeras Library

QKeras [10, 11] is an extension to Keras, a high-level API to define and train neural networks. It has been implemented to perform a drop-in replacement for certain layers of the model, related to weights and activation functions with a deep quantized version of Keras neural network.

QKeras is designed to remain a simple and consistent interface optimized for common functionalities in accordance with Keras design principles. For this purpose, the following set of layers have been implemented: Qconv2D, QActivation, QDense etc, to enable the conversion between non-quantized to quantized networks. To make your own quantization (QAT) it is needed to replace all variables and weights/bias created by Keras as well as output of arithmetic layers by quantized functions. Qactivation is used in both convolution (Qconv2D) and dense (Qdense) layers and acts at the end as a merging function for activation and quantization. For these layers, some parameters are interesting to mention.

Alpha is a parameter concerning the scale factor and should be applied before the activation function. This parameter by default is *None*. It can also indicate that the scale is computed as a floating-point number by the learning process. It can also force the scale to be an integral power of 2, which ends up for a hardware implementation in shifting the result of a convolution or dense layer to the right or left (positive shifts left, negative shifts right). For these practical reasons, in our experiments we opt for the latter setting.

Symmetric if set to, if set to *True,* ensures the trainable parameters to get the same maximum and minimum values after the clipping operation during quantization. The use of *stochastic_rounding* reveals to be useful in practice for improving accuracy. However, computing stochastic rounding might be quite heavy, so we set this parameter to *False.*

Table 4.4 describes the results obtained by our model after quantizing for different precisions. For example, the first convolution *q_conv2d* is set this way: bits=4, integer=0, symmetric=1. The 4-bit quantization of the entire model (weights, biases, and activations) achieves the better accuracy. When further reducing to 2 bits, the accuracy of the model decreases drastically.

Table 4.4　QKeras quantization for different precisions

Layer	Precision			Sparsity		
	4 bits	2 bits	Heterogeneous (4 bits, binary)	4 bits	2 bits	Heterogeneous (4 bits, binary)
q_conv2d W,b ReLU	(4,0,1) (4,1)	(2,0,1) (2,0)	(4,1,1), (4,1,1) (4,1)	0.1156	0.5131	0.1350
q_conv2d_1 W,b ReLU	(4,0,1) (4,1)	(2,0,1) (2,0)	binary, (4,1,1) (4,1)	0.1023	0.5341	-
q_conv2d_2 W,b ReLU	(4,0,1) (4,1)	(2,0,1) (2,0)	binary, (4,1,1) (4,1)	0.0985	0.5720	-
q_conv2d_3 W,b ReLU	(4,0,1) (4,1)	(2,0,1) (2,0)	binary, (4,1,1) (4,1)	0.1108	0.5483	-
q_conv2d_4 W,b ReLU	(4,0,1) (4,1)	(2,0,1) (2,0)	binary, (4,1,1) (4,1)	0.1973	0.5330	-
q_conv2d_5 W,b ReLU	(4,0,1) (4,1)	(2,0,1) (2,0)	binary, (4,1,1) (4,1)	0.2243	0.6550	-
q_conv2d_6 W,b ReLU	(4,0,1) (4,1)	(2,0,1) (2,0)	binary, (4,1,1) (4,1)	0.0975	0.5445	-
q_conv2d_7 W,b ReLU	(4,0,1) (4,1)	(2,0,1) (2,0)	binary, (4,1,1) (4,1)	0.1389	0.5975	-
q_conv2d_8 W,b ReLU	(4,0,1) (4,1)	(2,0,1) (2,0)	binary, (4,1,1) (4,1)	0.1754	0.5540	-
q_conv2d_9 W,b ReLU	(4,0,1) (4,1)	(2,0,1) (2,0)	binary, (4,1,1) (4,1)	0.2935	0.6840	-
q_dense W,b ReLU	(4,0,1) (4,1)	(2,0,1) (2,0)	(4,1,1), (4,1,1) (4,1)	0.5152	0.8756	0.4806
	Model Performance			**Total Sparsity**		
Accuracy Loss	96　% 0.148	76.1　% 0.747	94.6 % 0.163	0.4397	0.81	0.3318

The table shows, as an additional information, the model sparsity for various quantization scenario. This is a valuable metric to compare and see the trade-off between the accuracy and the computational cost of the model. The weights sparsity plays the role in reducing the number of calculations during the inference. When the sparsity is the same, the level of FLOPs remains constant. When the sparsity is too important (2 bits precision), the quantization becomes less effective, and the accuracy of the model is reduced. The sparsity for the last fully connected layer is like the pruning technique, where synapses between neurons are reduced.

These reasons led us to search for heterogeneous quantization, the trick being to find the right trade-off between accuracy requirements and hardware performance.

From a practical point of view, for the weights of the intermediate layers, we opted for an extremely low-bit quantization (we used *binary* quantization). The first and last classification layers were quantized to 4 bits, as well as the biases of the entire network. The activations of each quantized layer play an equally important role, so these neurons have not decreased in number of bits, the precision is maintained 4 bits.

This last model was implemented on the small Zybo board. The precision for each quantized layer and the accuracy of the model are described Table 4 (heterogeneous quantization). This method enabled us to achieve an accuracy slightly lower than the performance of the 4-bit model, more precisely a rate of 94.6%.

4.5.2.2 Quantized model and Experimental Setup

The quantized network used in our experiment, targets the small board Zybo Z7010 and explores the advantages of low-bit quantization. The major advantage of binary precision is that the pre-trained weights of the model (1.9 MB) fit very well within the on-chip memory. To achieve high memory throughput and very lightweight control paths, our hardware implementation does not leverage weight sparsity or compression. The low resource usage of multipliers with binary weights also enable to use a larger bit width for activations (4 bits), keeping accuracy high. Each network layer is an independent hardware block with its own dedicated resources and implementation, which enables to optimize parallelism and memory usage on a per-layer basis. The entire network fits inside the FPGA.

The approach of this efficient neural network implementation is presented in [17]. The hardware architecture generated by this method presents a total of 40 layers, with the following type: Sliding Window Layer, Neuron Layer,

Table 4.5   FPGA performance and resource utilization

LUT (logic)	LUTRAM	Slice Registers	Block RAM	DSP cores
9030 / 17600	4830 / 6000	11796 / 35200	60 / 60	37 / 80
(51.3 %)	(80.50 %)	(33.51 %)	(100 %)	(46.25 %)

Table 4.6   Model perpormance on FPGA

Performance	Latency	Power (FPGA only)	Power (Entire Chip)
178 images/s	26 ms	0.24 W $\sim$ 134 mJ/ image	1.75 W $\sim$ 983 mJ/ image

ReLU Layer, MaxPooling Layer, and Fork and Cat layers for synchronization of the parallel branches in the *inception* part of the network.

The resource utilization and performance of our quantized network implementation, is described Table 4.5 .

The BRAM is used for read-only memories of weights in neuron layers. All quantized MAC operations (multiply-accumulate) in neuron layers are implemented in distributed logic with LUTs. The MAC operations of most layers have a 1 b operand, which reduce the multiplications to tiny AND operations. Only the last layer actually implements a 4 b multiplication (0.5% of all MACs). The DSP cores are only used for address calculations within Sliding Window Layers.

The power consumption number is the total power estimation performed by the Xilinx Vivado synthesis suite. The processor subsystem of the Zynq chip would actually be mostly idle, so we report power both for the whole chip and for the FPGA only. The performance at 150MHz is summarized in following Table 4.6.

4.6 Conclusion

A selection of edge-AI boards and some optimization techniques have enabled us to investigate the possibilities of achieving high performance on a low budget. With a deep CNN model defined for a classification task, the accuracy achieved on 8-bit operating systems is around 97%. The efficiency of each board depends on processing speed and RAM availability. In our experiments, we found that performance is more limited by memory usage than by the number of neurons. In addition, we show how performance and energy efficiency can be affected by the cost of each board. To these measurements, further experiments using binary operations were carried out to address the option of a more energy efficiency at the expense of slightly degraded accuracy (94.6%). To find a suitable model, we used a hybrid aware quantization and described the methods enabling the maintain of an acceptable accuracy.

By focusing on this type of optimization related to the memory usage, i.e., an appropriate number of weights and limited bit widths, we have shown that high-performance inference can be achieved very efficiently. More specifically, the energy efficiency and power consumption achieved by each evaluation board is summarized as follows:

- Coral TPU: 3.12 mJ/image or 320 images/s/W
- STM32MP1: 232 mJ/image or 4.7 images/s/W
- Jetson GPU: 22.7 mJ/image or 44 image/s/W
- Zybo Z-7010: 983 mJ/image or 101.7 image/s/W

For further work, we plan to try out other optimization techniques linked to specific applications, for which these methodologies are of the utmost interest.

Acknowledgements

This work was supported by Key Digital Technologies Joint Undertaking (KDT JU) in EdgeAI "Edge AI Technologies for Optimised Performance Embedded Processing" project, grant agreement No 101097300.

References

[1] Claudionor N. Coelho Jr, Aki Kuusela, Shan Li, Hao Zhuang, Jennifer Ngadiuba, Thea Klacboe Aarrestad, Vladimir Loncar, Maurizio Pierini, Adrian Alan Pol, Sioni Summers, "Automatic heterogeneous quantization of deep neural networks for low-latency inference on the edge for particle detectors", Nature Machine Intelligence (2021)

[2] Zhang, Xiangyu & Zou, Jianhua & He, Kaiming & Sun, Jian. (2015). Accelerating Very Deep Convolutional Networks for Classification and Detection. IEEE Transactions on Pattern Analysis and Machine Intelligence.

[3] Tamara G. Kolda and Brett W. Bader. 2009. Tensor Decompositions and Applications. SIAM Rev, 51, 3 (August 2009), 455-500.

[4] T. Moreau *et al.,* "A Hardware-Software Blueprint for Flexible Deep Learning Specialization," in *IEEE Micro*, vol. 39, no. 5, pp. 8-16, 1 Sept.Oct. 2019, doi: 10.1109/MM.2019.2928962.

[5] Norman P. Jouppi, et al. 2017. In-Datacenter Performance Analysis of a Tensor Processing Unit. SIGARCH Comput. Archit. News 45, 2 (May 2017), 1-12.

[6] Vasilache, Nicolas, et al. "Tensor comprehensions: Framework-agnostic high-performance machine learning abstractions.", 2018.

[7] R. Zhao *et al.,* "Hardware Compilation of Deep Neural Networks: An Overview," *2018 IEEE 29th International Conference on Application-specific Systems, Architectures and Processors (ASAP)*, Milan, Italy, 2018, pp. 1-8, doi: 10.1109/ASAP.2018.8445088.

[8] Zhang, Tao *et al.* "cuTensor-Tubal: Efficient Primitives for Tubal-Rank Tensor Learning Operations on GPUs." IEEE Transactions on Parallel and Distributed Systems 31 (2020): 595-610.

[9] Pinzari, Ana et al. (2023). Power Optimized Wafer map Classification for Semiconductor Process Monitoring.

[10] Moons, Bert, et al. "Minimum energy quantized neural networks." *2017 51st Asilomar Conference on Signals, Systems, and Computers*. IEEE, 2017.

[11] Zhou, Shuchang, *et al.* "Dorefa-net: Training low bitwidth convolutional neural networks with low bitwidth gradients." *arXiv preprint arXiv:1606.06160* (2016).

[12] G. Dundar and K. Rose, "The effects of quantization on multilayer neural networks," in IEEE Transactions on Neural Networks, vol. 6, no. 6, pp. 1446-1451, Nov. 1995.

[13] B. G. Hoskins, M. R. Haskard and G. R. Curkowicz, "A VLSI implementation of multi-layer neural network with ternary activation functions and limited integer weights," Proceedings of International Conference on Microelectronics, Nis, Serbia, 1995, pp. 843-846 vol.2.

[14] R. Andri, L. Cavigelli, D. Rossi and L. Benini, "YodaNN: An Ultra-Low Power Convolutional Neural Network Accelerator Based on Binary Weights," 2016 IEEE Computer Society Annual Symposium on VLSI (ISVLSI), Pittsburgh, PA, USA, 2016, pp. 236-241

[15] Umuroglu, Yaman & Fraser, Nicholas & Gambardella, Giulio & Blott, Michaela & Leong, Philip & Jahre, Magnus & Vissers, Kees. (2017). FINN: A Framework for Fast, Scalable Binarized Neural Network Inference.

[16] Ritchie Zhao, et al. 2017. Accelerating Binarized Convolutional Neural Networks with Software-Programmable FPGAs. In Proceedings of the 2017 ACM/SIGDA International Symposium on Field-Programmable Gate Arrays (FPGA '17). Association for Computing Machinery, New York, NY, USA.

[17] Adrien Prost-Boucle, Alban Bourge, and Frédéric Pétrot. 2018. High-Efficiency Convolutional Ternary Neural Networks with Custom Adder Trees and Weight Compression. ACM Trans. Reconfigurable Technol. Syst. 11, 3, Article 15 (September 2018).

[18] N. P. Jouppi, et al. "Ten Lessons From Three Generations Shaped Google's TPUv4i: Industrial Product," 2021 ACM/IEEE 48th Annual International Symposium on Computer Architecture (ISCA), Valencia, Spain, 2021, pp. 1-14.

[19] Jacob, Benoit, et al. "Quantization and training of neural networks for efficient integer-arithmetic-only inference." Proceedings of the IEEE conference on computer vision and pattern recognition. 2018.

[20] Courbariaux, Matthieu, Bengio, Yoshua, et David, Jean-Pierre. Binaryconnect: Training deep neural networks with binary weights during propagations. Advances in neural information processing systems, 2015 , vol. 28 .

[21] Fraser, Nicholas J., et al. "Sealing binarized neural networks on reconfigurable logic." Proceedings of the 8th Workshop and 6th Workshop on Parallel Programming and Run-Time Management Techniques for Many-core Architectures and Design Tools and Architectures for Multicore Embedded Computing Platforms. 2017.

[22] Umuroglu, Yaman, et al. "Finn: A framework for fast, scalable binarized neural network inference." Proceedings of the 2017 ACM/SIGDA international symposium on field-programmable gate arrays. 2017.

[23] A. D. Vita, D. Pau, L. D. Benedetto, A. Rubino, F. Pĩl'trot and G. D. Licciardo, "Low Power Tiny Binary Neural Network with improved accuracy in Human Recognition Systems," 2020 23rd Euromicro Conference on Digital System Design (DSD), Kranj, Slovenia, 2020, pp. 309-315.

5

Designing Lightweight CNN for Images: Architectural Components and Techniques

Lilian Hollard, Lucas Mohimont, and Luiz Angelo Steffenel

Université de Reims Champagne-Ardenne, France

Abstract

While neural networks have brought about impressive advancements in computer vision tasks, these achievements heavily depend on computationally demanding resources, restricting their deployment. The decentralized paradigm of Edge AI computing aims to bring decisional capabilities directly to the edge, facilitating real-time decision-making, streamlined data processing, and reduced dependence on network connectivity. In some cases, it is possible to rely on cloud computing to offload processing tasks, but this can introduce latency issues that affect system responsiveness, security, and efficiency. Instead, searching for optimized neural networks for edge device deployment may lead to a better balance between computational efficiency and accurate analysis, empowering sensors to execute their roles effectively with minimal reliance on external resources. This paper reviews the landscape of deep learning architecture optimization tailored for edge devices. Within this survey, we delve into the state-of-the-art advancements in computer vision techniques optimized for edge computing. The challenges deploying and optimizing computer vision models on edge devices emphasize the importance of efficient computation and resource management while navigating the trade-offs between model performance and hardware constraints.

Keywords: neural network architecture, Edge AI, deep learning, neural architecture search, transformers, Edge vision, computer vision.

DOI: 10.1201/9788770041027-5   105

5.1 Introduction and Background

Neural networks enabled significant advancements in computer vision. However, these achievements often rely on computationally expensive resources, limiting deployment on less powerful devices. Despite the rapid adoption of cloud-based processing and cloud AI over the last decade, such offloading brings several inconveniences such as latency, bandwidth limitations, and security concerns. These challenges led to the development of Edge AI, which stands to the AI landscape. Edge Computing offers notable advantages, such as data ownership, heightened security, reduced latency, and decreased power consumption attributed to minimized back-and-forth communication with the cloud.

The term "Edge" encompasses a wide spectrum of devices and applications, including peripheral data centres (cloudlets [1], fog [2]) and IoT endpoints. Hence, it is not uncommon to partition the Edge according to the capabilities of the devices or the distance to end users. For example, the EdgeAI European project classifies edge solutions in three levels: Meta, Deep and Micro-Edge. Micro Edge represents mostly the final sensors and devices, Deep Edge lies in the vicinity (gateways, network routers) and Meta-Edge interfaces the Edge with external technologies such as the cloud. As a result, the size and capability of devices are determinant factors that differentiate various "Edge" application areas within Edge AI.

Bringing computer vision tasks to Micro-Edge devices such as microcontrollers is often complex due to resource limitations and computational constraints. These devices may struggle with the intensive processing demands of computer vision algorithms, making it difficult to perform analysis and decision-making directly at the edge. Offloading computer-vision tasks to "upper" layers is also a problem, as the data volume to be transmitted is far over more traditional IoT sensor data. Instead, Edge-AI computer vision requires optimized solutions adapted to the resource constraints of edge devices.

In this chapter, we review the most recent advances in computer vision methodologies for edge computing, with a specific emphasis on model architecture. While various established techniques such as quantization, pruning, and hardware optimization have been extensively investigated, our primary focus is the substantial enhancements that deep learning model architecture has witnessed over the last few decades. These enhancements have notably contributed to the improvement of Edge-AI. We explore the challenges faced in deploying and optimizing computer vision models on edge devices, the need for efficient computation and resource management, and the trade-offs between model performance and hardware constraints.

Furthermore, we run our own benchmarks to obtain uniform comparison results. Indeed, the deployment of deep learning models must emphasize model efficiency and comparison across various parameters. Metrics such as inference time, latency, training and inference costs, and other established indicators are crucial for researchers to demonstrate the contribution of new deep-learning techniques. However, researchers often assume a correlation among these metrics and report only a few of them, leading to partial conclusions and incomplete evaluations of different models [3].

Considering the types of models, different computational aspects may yield varying results. One example of bias in deep learning model optimization is relying solely on parameter-matched comparisons as a single metric, which may result in a flawed understanding of overall model performance. Shift-based convolution for instance, improves overall accuracy by offering a parameter-free alternative to traditional convolution but increases processing times. Memory access costs on different platforms or overall, unsatisfactorily optimized multi-branch model architecture for parallel computing might as well influence speeds metrics [3][7][8]. Therefore, models should evaluate multiple metrics on the targeted platform, as memory access and model parallelization are architecture dependent.

For the sake of reproducibility, most benchmarks presented in this chapter originate from the PyTorch Torchvision module benchmark. Our intention is to enable readers to replicate the results, although some models may exhibit slight variations from their original paper.

The remainder of the chapter is structured as follows: Section 1.2 explores the latest advancements in the computer vision research community using convolutional neural networks. Section 1.3 examines how Transformers revolutionized computer vision and how these techniques can be employed to reduce the overall computational cost. Section 1.4 investigate ConvNeXt convolution and its potential to elevate CNN models, enabling them to rival Transformers, while also exploring their utility in Edge computation. Section 1.5 covers the neural architecture search in an efficient computation scenario. We conclude this paper in Section 1.6, presenting some final remarks and research directions.

5.2 CNNs

Convolutional Neural Networks (CNN) are a class of deep learning models that excel when processing and analysing visual data. CNN enhanced the ability to learn intricate patterns and features from massive datasets, empowering deep learning to achieve remarkable breakthroughs in diverse areas, including computer vision but also natural language processing, speech recognition, recommendation systems and many more. The scalability, adaptability, and robustness of CNNs make it a dominant force in the breakout of AI technologies.

5.2.1 The pioneers

Optimising CNNs for low-power applications often involves weight pruning, quantisation, and model compression techniques. In theory, these methods position CNNs as ideal solutions for edge devices operating in energy-constrained environments; however, although essential for edge devices, these techniques often reduce accuracy. We cannot deny that the computer vision community made substantial progress since the remarkable performance of AlexNet's [9] first publication. Through architectural changes and optimisations, significant performance improvements extended several times state-of-the-art computer vision models while reducing computational demands.

ResNet [10] revolutionised the landscape of neural networks by introducing the concept of residual connections, a breakthrough that facilitated the construction of exceptionally deep models. This innovation proved instrumental in optimising the training of deeper layers. As a result, these extended CNN architectures attained unparalleled performance across diverse benchmark tasks.

ResNet's pioneering influence shaped the field of Deep Learning, serving as a cornerstone that inspired the architectural design of countless contemporary models. Residual connections provide an alternative pathway for the gradient to flow during backpropagation to address the vanishing gradient problem, as illustrated in Figure 5.1.

However, while many high-performing CNN models characterised by substantial numbers of parameters and FLOPS (Floating point operations per second) achieved impressive performance, the realm of Lightweight CNNs emerged as a potent contender. These efficient architectures, including EfficientNet [11][13], MobileNets [14][15], ShuffleNet [16][17], SqueezeNet [18], and ESPNet [19][20], devoted remarkable efforts to optimise CNNs by

ResNet Block

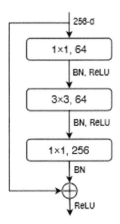

Figure 5.1 ResNet architecture [11]

renouncing the need for excessively deep and densely interconnected structures, aligning with the philosophy that ResNet and VGG [21] established.
The success of these Lightweight CNNs highlights their ability to achieve
competitive performance while maintaining a judicious balance between
model complexity, parameters, and FLOPS.

CNNs are widely utilised across various domains, such as classification,
object detection, segmentation, and other tasks. Currently, the object detection and segmentation research communities closely collaborate to enhance
classification CNNs and the other way around, recognising their pivotal role
as the backbone for object detection and segmentation models. Hence, the
forthcoming sections will not specifically address these distinctions, as they
all contribute to improving EdgeAI capabilities.

5.2.2 YOLO, first step towards fast object detectors

YOLO (You Only Look Once) [22], a deep learning model introduced in
2016, revolutionised object detection by providing real-time detection and
accurate results. Before YOLO, most object detection algorithms followed
a two-step approach, which was time-consuming and limited in case of
detection speed. On the other hand, YOLO formulated object detection as
a regression problem, dividing the input image into a grid and predicting
bounding boxes and class probabilities directly from the grid.

Newer versions of YOLO [23][24][25][26][27][28][29][30] incorporated concepts like anchor boxes, feature pyramid networks, and advanced network architectures (e.g., Darknet, ResNet) to improve detection accuracy and handle objects of various sizes. They also introduced multi-scale predictions, enabling detection at different resolutions within the network. However, these advanced network architectures, such as YOLOv5, have high memory requirements and need relatively powerful edge devices like the Jetson Nano with Nvidia GPU.

To circumvent such requirements, the object detection research community has three major perspectives. First, the improvement of YOLO-based models, both in terms of complex computation (e.g., YOLOv7 [25]) and adapting YOLO for edge computation (e.g., TinyssimiYOLO [31], YOLOv3 Tiny[26], YOLOv5 Nano [23], YoloNAS [32]). Second, the enhancement of backbones for object detection, such as combining MobileNetV3 [33] with SSD320 [34] detection heads. One last approach is the exploration of Transformers-based object detectors.

The miniaturisation of YOLO models is still ongoing research. Recent efforts have focused on reducing the architecture to create highly flexible, memory-efficient, and ultra-lightweight object detection networks with less than 0.5MB of memory. However, these optimised models are most suitable for detecting a few classes. For example, TinyissimiYOLO [31] performs well for up to three classes, and challenges remain when trying to improve the edge-oriented benchmark on datasets like MS COCO [35], which consists of 80 classes. As demonstrated in TinyissimoYOLO, a plain-architecture model still exhibits great potential for efficient inference on microcontrollers or edge devices.

In addition to YOLO object detectors, there is a significant effort to enhance classifier CNNs, which extend to backbone CNNs for object detector models in a broader context. These CNN architectures undergo rigorous benchmarking on datasets like ImageNet but also on datasets such as MSCOCO and Pascal VOC to address object detection and segmentation tasks. Indeed, the emergence of SSD detector heads and Mask R-CNN segmentation heads catalysed a distinct research avenue, prompting a concentrated exploration of classification models or backbone designs specifically tailored for advancing object detection capabilities.

Research to enhance classifiers changed the recent panel of YOLO models. Since YOLOv4 [30], the composition of YOLO models depends on CSPNet [36] block modifications, an architecture that already enabled known architectures such as ResNet, ResNeXt, and DenseNet to reduce

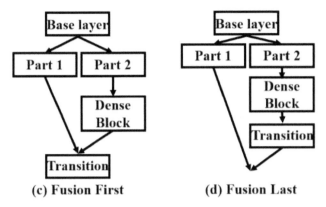

Figure 5.2 CSPNet (Identity Block - DenseNet)

computational cost while preserving accuracy. It effectively reduces computational bottlenecks (YOLOv3's computational bottlenecks can be reduced by 80%) and memory costs. ResNeXt already proved that cardinality can be more efficient than width and height. CSPNet divides feature maps into two main parts: one used to create an identity block (DenseNet, ResNet, MobileNet, etc.) and the other that is combined at the output after or before a transition layer, as shown in Figure 5.2.

5.2.3 Convolutional Neural Network architecture improvements

MobileNetv1 was one of the first CNN architectures specifically created to bring efficiency to mobile and embedded vision applications. Its main improvement was the efficient use of depthwise separable convolutions to build lightweight neural networks. MobileNet was nearly as accurate as VGG16, with 32 times less size and 27 times less computation. MobileNetv1 performance on the ImageNet dataset achieved a top-1 of 68.4%. MobileNetv2, on the other hand, improved MobileNetv1 drastically while preserving the same mobile-first philosophy, using inverted residual block and linear bottlenecks. MobileNetv2's linear bottleneck does not incorporate linear activation within its narrow input and output layer. Instead, it incorporates non-linearity after each expanded layer of the bottleneck.

The hypothesis of MobileNetv2 stated that ReLU can preserve complete information only if the i-th feature input lies in a low-dimensional subspace of the input space. Researchers showed through experimental evidence that using non-linear layers in the input/output of bottlenecks impacts the

(a) Residual block   (b) Inverted residual block

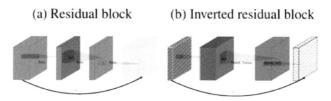

Figure 5.3 MobileNetv2 block

model's performance by several percent. Figure 5.3 illustrates the difference between non-linearity in the residual and inverted residual blocks. Residual block architecture skip connections with fewer feature maps between connections. In contrast, an inverted residual block broke this relation by using an expansion of feature maps. As a result, MobileNetV2 TOP-1 ImageNet performance reaches 71.978%, with only 2.6M parameters and 0.3G Flops. Table 5.1 details the performance of different MobileNet models when conducting the ImageNet classification.

Since MobileNet, a vast majority of modern networks adopted depth-wise separable convolutions. ShuffleNetv1 and v2 introduced practical guidelines for efficient network design, resolving some MobileNet issues.

ShuffleNet v2 stated that the expensive use of depth-wise separable convolutions and grouped convolution increase memory access cost. Also, element-wise operations have a high MAC (Memory Access Cost) and FLOP cost, even with a small parameter count. The ShuffleNet architecture thus introduces an architecture using balanced convolutions of equal channel width, reducing the degree of fragmentation and reduced element-wise operations, surpassing MobileNetv2 with an ImageNet top-1 performance of

Table 5.1 CNNs based model optimization since AlexNet

Model	ImageNet Top 1(%)	Parameters (M)	FLOPs (G)
ShuffleNetv2x0.5	60.5	1.3	0.04
ShuffleNetv2x1.0	69.36	2.27	0.14
MobileNetv3 Small	67.4	2.5	0.06
MobileNetv2	71.97	2.6	0.3
ShuffleNetv2x1.5	72.99	3.5	0.30
ShuffletNetv2x2.0	76.23	7.3	0.58
AlexNet	56.52	61	0.71
VGG16	76.3	132.8	7.61

72.99%, with comparable FLOPs of 0.3G and 3.5M parameters. Table 5.1 benchmarks ShuffletNet v2 across various scales.

5.2.4 Tackling memory consumption

Memory consumption is a significant concern when optimizing CNNs for mobile computing applications. State-of-the-art models for mobile and edge often employ grouped and depth-wise convolutions to reduce overall model parameters [37][14][15][39][16][17]. However, these models require more computation time and memory per layer, which may pose challenges for edge AI models focused on video stream processing.

Therefore, in addition to the architecture optimization research of CNNs, researchers made significant efforts to achieve extreme memory consumption optimization for microcontroller use cases. A remarkable example of this is MCUNet [40], which demonstrated impressive potential for memory optimization by simply enhancing the memory workflow of CNNs following the MobileNet architecture philosophy previously mentioned.

MCUNet significantly reduces memory usage for MobileNetv2, fitting it within a mere 320kB of RAM. This impressive feat is accomplished through two key strategies: first, by identifying the optimal input resolution size and adapting the model width to achieve the most efficient neural architecture size. Second, it leverages the characteristic of depth-wise convolutions, which do not perform filtering across channels, allowing each channel to be computed in a temporary buffer. This approach substantially reduces overall memory consumption, computing the input and the output feature map as one shared memory, with one additional buffer to compute and transfer the data. MCUNetv2 [41] goes further by optimizing memory usage through patch-based computation. Instead of processing the entire feature width and height, it strategically employs small input portions to generate activation maps, leading to more efficient memory utilization. Table 5.2 describes MobileNets memory consumption improvement with MCUNets.

5.2.5 Structural re-parameterization

Within this survey, we showcase research aimed at enhancing the architecture of models commonly referred to as "mobile". However, these mobile-first models rely heavily on grouped and depth-wise convolutions, which induce many other computational challenges.

Table 5.2   MCUNet memory optimization compared to MobileNet and MobileNetv2

Model	ImageNet Top 1 (%)	SRAM
MobileNetv1	68.4	NS
MobileNetv2	69.8	1.8 MB
MCUNetv1-int8	60.3	238 kB
MCUNetv2-int8	64.90	**196 kB**
MCUNetv1-int8	68.5	452 kB
MCUNetv2-int8	**71.8**	465 kB

Grouped and depth-wise convolutions utilize 1x1 convolutions not well-optimized for certain architectures. In contrast, 3x3 convolution architectures are more efficient on generic GPUs than 1x1 convolutions. Multi-branch design models like ResNet [7] or branch-concatenation in Inception [42] encounter similar issues, making them less efficient for parallel architectures like GPUs due to additional overhead, such as kernel launching and synchronization. Residual connections also face challenges in retaining convoluted feature maps in memory during the computation of multi-branch deep learning architectures.

To address these challenges, recent research suggests structural re-parameterization to revert to early deep learning plain models like VGG [21], Darknet [22], and AlexNet [9], which are theoretically efficient for edge computation. However, these models no longer compete in terms of accuracy and overall performance with the current state-of-the-art models.

Residual connections and multi-branch architecture [43][44][42][36][39] [45][46], to cite only a few, are indeed essential components in deep learning architectures. Their introduction addresses the vanishing gradient problem, which occurs when gradients diminish as they propagate through deep networks during training.

It is challenging for a plain model to achieve comparable performance to a multi-branch architecture. The complex structure of multi-branch architectures often slows down inference, as the combination of small operators is not favourable for devices with strong parallel computing capabilities like GPUs. Taking a more edge-centric perspective, utilising multi-branch structures necessitates significant cache memory, as these structures demand the model to retain the feature maps of each branch in memory before processing to the subsequent layer. However, the benefits of multi-branch architecture mainly apply while training [47][7][46].

Structural re-parameterization involves transferring the knowledge gained from multi-branch architecture during training into a single plain convolution block for inference.

MobileOne TOP-1 ImageNet performance on multiple scaling is 71.4% and 75.9% for MobileOne-S0 and MobileOne-S1 respectively, both under 1G Flops. Table 5.2 lists the complete results. Other re-parametrization models, like RepVGG, accompany the comprehensive results in Table 5.3.

While the training cost may be significant, improving performance at the cost of additional training resources is acceptable if the deployed model fits the size and computing power required for edge devices. Hence, one

Table 5.3 CNNs based model optimization since AlexNet

Model	ImageNet Top 1(%)	Parameters (M)	FLOPs (G)
AlexNet	56.52	61	0.71
EfficientNet B0	77.69	5.2	0.32
EfficientNet B1	78.64	7.79	0.69
EfficientNet B2	80.6	9.1	1.09
EfficientNet B3	82.2	12.2	1.83
EfficientNet B4	83.4	19.34	4.38
EfficientNet B7	84.122	66.34	37.75
EfficientNetv2 Large	85.808	118.5	56.08
EfficientNetv2 Medium	85.112	54.1	24.58
EfficientNetv2 Small	84.228	21.4	8.37
ESPNetv2	72.1	3.49	0.28
MobileNetv1	68.4	2.6	NS
MobileNetv2	71.97	2.6	0.3
MobileNetv3 Small	67.4	2.5	0.06
MobileOne-S0	71.4	2.1	0.275
MobileOne-S1	75.9	4.8	0.825
MobileOne-S2	77.4	7.8	1.29
MobileOne-S3	78.1	10.1	1.89
MobileOne-S4	79.4	14.8	2.97
RepVGG-A0	72.4	8.3	1.4
RepVGG-A1	74.5	12.8	2.4
RepVGG-B0	75.1	14.3	3.1
ResNet101	77.374	44.5	7.80
ResNet152	78.312	60.19	11.51
ResNet18	69.75	11.6	1.81
ResNet34	73.314	21.7	3.66
ResNet50	76.13	25.5	4.09
ShuffleNetv2x0.5	60.5	1.3	0.04
ShuffleNetv2x1.0	69.36	2.27	0.14
ShuffleNetv2x1.5	72.99	3.5	0.30
ShuffletNetv2x2.0	76.23	7.3	0.58
SqueezeNet	57.5	1.2	0.35
VGG16	76.3	132.8	7.61

can use a high-end GPU server for training before deployment on an edge device.

5.3 Transformers for EdgeAI

Initially proposed for natural language processing tasks, transformers have also found exploration in computer vision [48][49][50][51][52][53]. The Vision Transformer (ViT) [52], introduced in 2020, adapted the Transformer architecture specifically for computer vision by replacing CNN-based back-bones with Transformer encoders. ViT achieved competitive performance on image classification benchmarks, indicating the effectiveness of Transformers in vision tasks. However, Transformers require substantial amounts of data to work well, and ViT struggles to perform on ImageNet-1K 0, which already contains over one million images. Convolution still plays an important role in Transformers.

5.3.1 Hybrid transformers

Research towards the miniaturization of Transformers focuses on utilizing CNNs as feature extractors with low computational cost. Models like MobileViT and its latest versions [37][50][49] achieved impressive performance with significantly fewer parameters and floating-point operations than previously published computer vision Transformers. The concept involves incorporating multiple CNN blocks within transformers to enhance the extraction of features, as illustrated in Figure 5.4. While transformers like ViT demonstrated remarkable capabilities in natural language processing tasks, they often lack the reliance on powerful feature extractors, such as CNNs, which may explain the challenges faced by such models in achieving efficient training with limited data, as seen in the struggle encountered

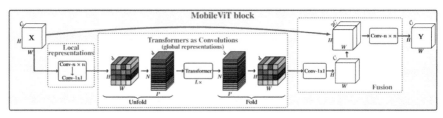

Figure 5.4   MobileViT block [37].

when training ViT based network on datasets like ImageNet1K. These contributions may lead to new backbones for object detection, segmentation, and classification in the computer vision community, utilizing these hybrid models.

As mentioned earlier, the number of parameters is not the sole metric for comparing results; a compelling example is the comparison between MobileNetv2 and its supposed enhancement, MobileViT. MobileViT fails to match the computing speed of MobileNetv2 on the iPhone 12 neural engine, as the latter achieves an inference time of 0.9ms, while MobileViT requires 7.28ms per inference despite having a similar number of parameters.

Transformers inherently tend to be slower than CNNs for several reasons. First, Vision Transformers (ViTs) are designed to leverage dedicated CUDA kernels on GPUs, enabling improved scalability and efficiency. In contrast, CNNs benefit from device-level optimizations like batch normalization fusion with convolutional layers, 3x3 convolution optimization, and other techniques. This observation suggests there is still room for improvement in optimizing transformers at a lower computational level. Despite their potential, transformers must continue to evolve to achieve faster and more efficient performance comparable to that of CNNs in specific contexts.

MobileViTv2's [50] enhancement primarily focused on optimizing the self-attention operation within transformers. As previously mentioned, researchers have a significant opportunity to improve attention layers like Multi-Head Attention (MHA). The computational complexity of the MHA layer is typically $O(k^2)$ whereas MobileViTv2's version of MHA has reduced it to $O(k)$ through the implementation of element-wise operations. The concept involves using element-wise operations such as summation, multiplication, and softmax instead of more computationally intensive operations like batch-wise matrix multiplication, which is quadratically expensive.

Previous efforts to optimize self-attention, such as the Linformer [37] approach, decompose the self-attention operation into smaller segments via linear projections, effectively reducing the complexity from $O(k^2)$ to $O(k)$. However, Linformer still employs resource-intensive operations to learn global representations within MHA, which could pose challenges for deploying such models on devices with limited resources. Other methods have managed to reduce complexity to $O(k)$ but often suffer substantial performance degradation.

In contrast, MobileViTv2 outperformed MobileViTv1 by approximately 1% and exhibited a significant speed boost, running 3.2 times faster on comparable devices. This advancement underscores the potential of optimizing self-attention operations within transformers while maintaining robust performance, especially in constrained computing environments.

In recent developments, there has been progress in enhancing hybrid architectures combining CNN and Transformers for mobile devices. Mobile-ViTv3 [49] emerged as an improved iteration of the initial MobileViT architecture. This advancement involves substituting resource-intensive 3x3 convolutional layers with more efficient depthwise and 1x1 convolutions. Additionally, the integration of residual connections contributed to an overall performance boost for the MobileViT v1 design. Furthermore, this enhancement opened avenues for scaling the width of the MobileViTv3 model. The removal of the costly 3x3 convolutions led to a reduction in parameters and FLOPs, resulting in improved scalability while maintaining or enhancing performance.

Table 5.4 gives comprehensive results around each MobileViT version compared to Swin-T as an accuracy gap to achieve for the transformer-based model. While Swin Transformer [51] and DETR [48] significantly improved tackling the ImageNet classification and MS COCO challenges, they remain less suitable for edge computing due to their computational demands. Nevertheless, the technological advancements achieved through the rise of transformers could be instrumental in enhancing convolutions.

Table 5.4   Optimized transformers and hybrid transformers performance by scale

Models	ImageNet Top 1 (%)	Parameters (M)	FLOPs (G)
MobileViT-XXS	69.4	1.3	**0.4**
MobileViTv2-0.5	70.2	1.4	0.5
MobileViTv3-XXS	70.98	1.2	0.28
MobileViTv3-0.5	72.33	1.4	0.48
MobileViTv2-1.0	78.1	4.9	1.8
MobileViTv3-XS	76.7	2.5	0.92
MobileViTv3-1.0	**79.64**	5.1	1.87
MobileViTv3-S	79.3	5.8	1.84
MobileViTv2-0.5	70.2	1.4	**0.5**
MobiletViTv2-2.0	81.2	18.5	7.5
Swin-T	**81.3**	28.3	4.5
MobileNetv2	**71.978**	2.6	0.3

Consequently, using the benefits of the previously mentioned miniaturization techniques with these technological advancements in an edge-computing context becomes a promising avenue for further progress.

5.4 ConvNeXts

While Transformers are a significant breakthrough in computer vision, recent advancements, such as the ConvNeXt [11] convolution, quietly surpassed their performance in computer vision. The foundation of ConvNeXt lies in the adaptation of ResNet, which serves as a starting point, leveraging techniques inspired by transformers to fill the gap between ResNet's and Swin Transformer's performance (Figure 5.5).

To achieve this, ConvNeXt introduces a new CNN design inspired by Transformers. Firstly, it employs the patchify technique, which involves flattening the input into a vector of smaller patches, an idea initially introduced in Vision Transformers (ViT) to exploit text-based transformer techniques for processing 2D images effectively. Secondly, the training recipe aligns closely with the strategies employed in Swin Transformers and DeiT.

ConvNeXt Block

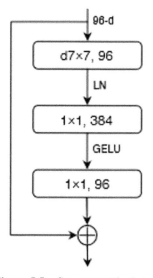

Figure 5.5 ConvNeXt Block [11]

Table 5.5   ConvNeXt compared to hybrid transformers performance by scale.

Models	ImageNet Top 1 (%)	Parameters (M)	FLOPs (G)
AlexNet	56.52	61	0.71
VGG16	76.3	132.8	7.61
ResNet152	78.312	60.19	11.51
EfficientNetv2 Large	85.808	118.5	56.08
MobiletViTv2-2.0	81.2	18.5	7.5
Swin-T	81.3	28.3	4.5
ConvNeXt-T	82.9	28.6	4.5
ConvNeXt-S	83.616	50.2	8.68
ConvNeXt-B	84.06	88.6	15.36
ConvNeXt-L	84.414	197	34.36

As a result, ConvNeXt exhibits less sensitivity to image shift-invariance. The training process is extensively enhanced through data augmentation, longer training epochs, and the AdamW optimizer, leading to improved results.

Furthermore, ConvNeXt introduces a compelling microdesign aspect, replacing the conventional ReLU activation function with GeLU (Gaussian Error Linear Unit), a different non-linear activation function. Additionally, the model incorporates fewer activations and norms than those found in transformer architecture (Recognizing the initial hypothesis involving non-linearity in MobileNetv2). These adaptations contribute to the model's overall efficiency and performance, effectively leveraging the power of transformer-inspired concepts within a CNN framework.

ConvNeXt's new philosophy for CNNs might open the path for new mobile architecture. The benchmark in Table 5.5 enlightens ConvNeXT performance compared to models with similar computing performance.

5.5 Neural Architecture Search

Amidst the collection of hand-crafted neural networks, the question arises: Can we venture into automatic network architecture design, to reduce the dependency on deep learning expert insights? The Neural Architecture Search (NAS) literature categorizes two primary domains: Evolutionary Algorithms (EA) and Reinforcement Learning (RL). Evolutionary algorithms utilize a pool of candidate architectures, each with its respective accuracy. Only a limited number of top-performing candidates evolve further. Should these evolved candidates exhibit enhanced accuracy, the candidate pool is accordingly updated. On the other hand, Reinforcement Learning (RL) employs an LSTM Agent to generate a string, serving as a dictionary of convolution operations to execute on hardware to train and test. The accuracy serves as the reward signal for this operation, and the LSTM Agent subsequently refines and produces another dictionary block.

The initial NASnet [54] model lacks consideration for runtime or computational efficiency. The search space for potential architectures is inherently resource intensive. While the LSTM Agent discovered architecture proves superior to manually crafted ones, it inherits the complexity identified in this survey as challenging for edge devices. This complexity arises from the substantial memory requirements due to the neural architecture search algorithm's reliance on a configuration of five cells per layer, each with three potential residual depth connections (from the previous cell's output, the cell before the previous one, and the previous block's output within the current cell).

Most NAS methods [54][55] explore architectural spaces to construct intricate cells, subsequently employing these cells with identical configurations throughout the network. Unfortunately, this approach lacks the potential for layer wise diversity. MnasNet [56], a breakthrough in Aware Neural Architecture Search for Mobile, introduces an edge computation model for inference by selecting a considerably smaller number of convolution blocks. This time, the LSTM agent selects hand-crafted architectures in alignment with the MobileNet philosophy. Employing such block-based designs reduces the search space from 10^{39} options to 10^{13}.

The search algorithm within MnasNet also introduces a multi-objective reward system that combines validation accuracy and a metric for real-world latency on mobile devices. This dual-objective approach optimization creates architectures that excel in both accuracy and efficiency for real-world performance.

5.5.1 NAS scale study

While NAS is primarily concerned with discovering novel convolutions based on accuracy-to-speed trade-offs, EfficientNet [12][13] takes a different approach, aiming to optimise a manually designed model by identifying the optimal balance among width, depth, and resolution (Figure 5.6). Although these components might seem independent, EfficientNet's case study highlights that achieving superior accuracy requires simultaneous optimization of all three components rather than considering them separately.

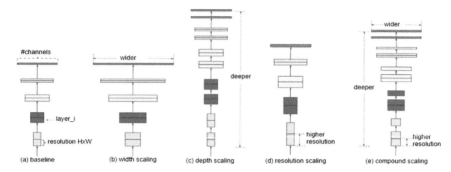

Figure 5.6   EfficientNet Scaling [12]

Table 5.6   Efficient neural network architectures with neural network search

Models	ImageNet Top 1(%)	Parameters (M)	FLOPs (G)
ResNet50	76.13	25.5	4.09
NASNet-A (4 @ 1056)	74	5.3	0.56
NASNet-B (4 @ 1536)	72.8	5.3	0.48
NASNet-C (3 @ 960)	72.5	4.9	0.558
MobileNetv1_efficientNetv1 (d=1.4,w=1.2,r=1.3)	75.6		2.3
MobileNetv2_efficientNetv1 (d=1.4, w=1.2, r=1.3)	77.4		1.3
PNASNet-5 (N = 3, F = 54)	74.2	5.1	0.58
PNASNet-5 (N = 4, F = 216)	82.9	86.1	25
EfficientNet_B0	77.692	5.2	0.32
EfficientNet_B1	78.642	7.79	0.69
EfficientNetv2_Small	84.228	21.4	8.37
EfficientNetv2_Medium	85.112	54.1	24.58
EfficientNetv2_Large	85.808	118.5	56.08
MobileNetv1	68.4	2.6	ND
MobileNetv2	71.978	2.6	0.3

To enhance accuracy and efficiency, the interplay between depth, width, and resolution is computed and then trained on the CIFAR dataset, allowing rapid assessment of the newly formulated model. Subsequently, this model is scaled up for evaluation in an ImageNet context. Building upon this foundation, EfficientNetv2 introduces a further innovation: the incorporation of fused MBConv (MobileNet-like convolution). This mechanism combines the 1x1 expansion convolution with the subsequent 3x3 depthwise convolution into a single 3x3 operation, streamlining both processes and enhancing overall efficiency. Table 5.6 compares the performance of

NAS-based models against a list of comparable and well-known hand-crafted architectures.

5.6 Conclusion

Efficient neural network architectures are a subject of ongoing intensive research within the deep learning community. This research aims to harness the scalability potential of Convolutional Neural Networks (CNNs) for emerging edge computing paradigms. Since the publication of AlexNet, these architectures not only improved accuracy but also advanced the state-of-the-art through optimised proposals. However, developing efficient neural networks for mobile and edge devices highlights the challenge of crafting such models manually.

Throughout this work, we presented a comprehensive array of optimised neural network architectures tailored for edge devices, encompassing more than just microcontrollers. While these architectures may exhibit discrepancies and contradictions, they collectively highlight the deep learning community's commitment to refining model architectures. This emphasises the absence of a one-size-fits-all architecture for edge devices and the necessity for benchmarks when searching for neural network architectures to fit our needs. This chapter provides researchers with a global perspective on significant advancements and their pros and cons, fostering a deeper understanding.

Acknowledgements

This research was conducted as part of the EdgeAI "Edge AI Technologies for Optimised Performance Embedded Processing" project, which has received funding from KDT JU under grant agreement No 101097300. The KDT JU receives support from the European Union's Horizon Europe research and innovation program and Austria, Belgium, France, Greece, Italy, Latvia, Luxembourg, Netherlands, and Norway.

References

[1] Buyya R., Yeo C. S., Venugopal, S., Broberg J., Brandic I. 'Cloud computing and emerging IT platforms: Vision, hype, and reality for delivering computing as the 5th utility', Future Generation Computer Systems, Volume 25, Issue 6, 2009, doi: 10.1016/j.future.2008.12.001.

[2] Steffenel L. A., "Improving the Performance of Fog Computing Through the Use of Data Locality," 2018 30th Int. Symposium on Computer Architecture and High Performance Computing (SBAC-PAD), Lyon, France, 2018, pp. 217-224, doi: 10.1109/CAHPC.2018.8645879.

[3] Mostafa Dehghani et al. The Efficiency Misnomer. The Tenth International Conference on Learning Representations (ICLR), April. 2022. doi: 10.48550/arXiv.2110.12894.

[4] Bichen Wu et al. "Shift: A Zero FLOP, Zero Parameter Alternative to Spatial Convolutions". In: 2018 IEEE/CVF Conference on Computer Vision and Pattern Recognition. Salt Lake City, UT: IEEE, June 2018, pp. 9127–9135. doi: 10.1109/CVPR.2018.00951.

[5] Andrew Brown, Pascal Mettes, and Marcel Worring. "4-Connected Shift Residual Networks". In: 2019 IEEE/CVF International Conference on Computer Vision Workshop (ICCVW). Seoul, Korea (South): IEEE, Oct. 2019, pp. 1990–1997. doi: 10.1109/ICCVW.2019.00248.

[6] He Y., Liu X., Zhong H., Ma Y., 'AddressNet: Shift-Based Primitives for Efficient Convolutional Neural Networks,' 2019 IEEE Winter Conference on Applications of Computer Vision (WACV), Waikoloa, HI, USA, 2019, pp. 1213-1222, doi: 10.1109/WACV.2019.00134.

[7] Ding X., Zhang X., Ma N., Han J., Ding G., Sun J. "RepVGG: Making VGG-style ConvNets Great Again". In: Proceedings of the IEEE/CVF conference on computer vision and pattern recognition. 2021. p. 13733-13742.

[8] Ma N., Zhang X., Zheng HT, Sun J. "ShuffleNet V2: Practical Guidelines for Efficient CNN Architecture Design", Proceedings of the European Conference on Computer Vision (ECCV), 2018, pp. 116-131.

[9] Krizhevsky A., Sutskever I., Hinton G.E. "ImageNet classification with deep convolutional neural networks". Communications of the ACM 60, 6 (June 2017), 84–90. doi: 10.1145/3065386.

[10] He K., Zhang X., Ren S., Sun J., "Deep Residual Learning for Image Recognition," 2016 IEEE Conference on Computer Vision and Pattern Recognition (CVPR), Las Vegas, NV, USA, 2016, pp. 770-778, doi: 10.1109/CVPR.2016.90.

[11] Liu Z., Mao H., Wu C.-Y., Feichtenhofer C., Darrell T., Xie S., "A ConvNet for the 2020s," 2022 IEEE/CVF Conference on Computer Vision and Pattern Recognition (CVPR), New Orleans, LA, USA, 2022, pp. 11966-11976, doi: 10.1109/CVPR52688.2022.01167.

[12] Tan, M., Le, Q. "EfficientNET: Rethinking model scaling for convolutional neural networks". In: International conference on machine learning (ICML). 2019. p. 6105-6114. url: http://arxiv.org/abs/1905.1 1946

[13] Tan, M., Le, Q. "EfficientNETv2: Smaller models and faster training". In: International conference on machine learning (ICML). 2021. p. 10096-10106. doi: 10.48550/arXiv.2104.00298.

[14] Howard A. G., Zhu M., et al. "Mobilenets: Efficient convolutional neural networks for mobile vision applications". arXiv preprint arXiv:1704.04861. 2017.

[15] Sandler M., Howard A., Zhu M., Zhmoginov A., Chen L.C. "MobileNetV2: Inverted residuals and linear bottlenecks". In: Proc. of the IEEE Conference on Computer Vision and Pattern Recognition (CVPR). 2018. p. 4510-4520.

[16] Zhang X., Zhou X., Lin M., Sun J., "ShuffleNet: An Extremely Efficient Convolutional Neural Network for Mobile Devices," 2018 IEEE/CVF Conference on Computer Vision and Pattern Recognition, Salt Lake City, UT, USA, 2018, pp. 6848-6856, doi: 10.1109/CVPR.2018.00716.

[17] Ma N., Zhang X., Zheng HT., Sun J. "ShuffleNet V2: Practical Guidelines for Efficient CNN Architecture Design". In: Computer Vision – ECCV 2018. LNCS, vol 11218. doi: 10.1007/978-3-030-01264-9_8.

[18] Iandola F.N., Han S., Moskewicz MW., Ashraf K., Dally WJ., Keutzer K. "SqueezeNet: AlexNet-level accuracy with 50x fewer parameters and <0.5MB model size". arXiv preprint, Nov. 2016. doi: 10.48550/arXiv.1602.07360.

[19] Mehta S., Rastegari M., Caspi A., Shapiro L., Hajishirzi H. "ESPNet: Efficient Spatial Pyramid of Dilated Convolutions for Semantic Segmentation". In: Computer Vision – ECCV 2018. LNCS vol 11214. doi: 10.1007/978-3-030-01249-6_34.

[20] Mehta S., Rastegari M., Shapiro L., Hajishirzi H., "ESPNetv2: A Light-Weight, Power Efficient, and General Purpose Convolutional Neural Network," 2019 IEEE/CVF Conference on Computer Vision and Pattern Recognition (CVPR), Long Beach, CA, USA, 2019, pp. 9182-9192, doi: 10.1109/CVPR.2019.00941.

[21] Simonyan K., Zisserman A. "Very Deep Convolutional Networks for Large-Scale Image Recognition". In: 3rd International Conference on Learning Representations (ICLR), San Diego, CA, USA, May 7-9, 2015. doi: 10.48550/arXiv.1409.1556.

[22] Redmon J., Divvala S., Girshick R., Farhadi A., "You Only Look Once: Unified, Real-Time Object Detection," 2016 IEEE Conference on Computer Vision and Pattern Recognition (CVPR), Las Vegas, NV, USA, 2016, pp. 779-788, doi: 10.1109/CVPR.2016.91.

[23] Glenn, J., " YOLOv5 by Ultralytics". doi: org/10.5281/zenodo.3908559.

[24] Chuyi Li et al. "YOLOv6: A Single-Stage Object Detection Framework for Industrial Applications". Sept. 2022. doi: 10.48550/arXiv.2209.02976.

[25] Wang CY., Bochkovskiy A., Liao HY. M., "YOLOv7: Trainable bag-of-freebies sets new state-of-the-art for real-time object detectors." Proceedings of the IEEE/CVF Conference on Computer Vision and Pattern Recognition. 2023.. doi: 10.48550/arXiv.2207.02696.

[26] Redmon J., Farhadi A. "YOLOv3: An Incremental Improvement". Apr. 2018. doi: 10.48550/arXiv.1804.02767.

[27] Yan W., Liu T., Fu Y., "YOLO-Tight: An Efficient Dynamic Compression Method for YOLO Object Detection Networks". In: 2021 13th International Conference on Machine Learning and Computing. ICMLC 2021. event-place: Shenzhen, China. New York, NY, USA: ACM, 2021, pp. 378–384. doi: 10.1145/3457682.3457740.

[28] Redmon J., Farhadi A., "YOLO9000: Better, Faster, Stronger," 2017 IEEE Conference on Computer Vision and Pattern Recognition (CVPR), Honolulu, HI, USA, 2017, pp. 6517-6525, doi: 10.1109/CVPR.2017.690.

[29] Fang F., Wang L., Ren P. "Tinier-YOLO: A Real-Time Object Detection Method for Constrained Environments". In: IEEE Access 8 (2020), pp. 1935–1944. doi: 10.1109/ACCESS.2019.2961959.

[30] Bochkovskiy A., Wang C.Y., Liao H-Y., M. "YOLOv4: Optimal Speed and Accuracy of Object Detection". Apr. 2020. doi: 10.48550/arXiv.2004.10934.

[31] Moosmann J., Giordan M., Vogt, C., Magno, M. "TinyissimoYOLO: A Quantized, Low-Memory Footprint, TinyML Object Detection Network for Low Power Microcontrollers". (2023). arXiv preprint, url:http://arxiv.org/abs/2306.00001

[32] Aharon S., Louis-Dupont, Ofri Masad, Yurkova K., Lotem F., Lkdci, Khvedchenya E., Rubin R., Bagrov N., Tymchenko B., Keren T., Zhilko A., Eran-Deci. "Super-Gradients". 2021. doi: 10.5281/ZENODO.7789328.

[33] Howard A., Sandler M., Chu G., et al. "Searching for MobileNetV3". Proceedings of the IEEE International Conference on Computer

Vision, Seoul, 27 October-2 November 2019, 1314-1324. doi: 10.1109/ICCV.2019.00140

[34] Liu W., Anguelov D., Erhan D., Szegedy C., Reed S., Fu C-Y., Berg A.C. "SSD: Single Shot MultiBox Detector". In: Computer Vision – ECCV 2016. LNCS vol 9905. doi: 10.1007/978-3-319-46448-0_2

[35] Lin, TY. et al. "Microsoft COCO: Common Objects in Context". In: ECCV 2014. LNCS vol 8693. doi: 10.1007/978-3-319-10602-1_48

[36] Wang C.-Y., Mark Liao H.-Y., Wu Y.-H., Chen P.-Y., Hsieh J.-W., Yeh I.-H., "CSPNet: A New Backbone that can Enhance Learning Capability of CNN," 2020 IEEE/CVF Conference on Computer Vision and Pattern Recognition Workshops (CVPRW), Seattle, WA, USA, 2020, pp. 1571-1580, doi: 10.1109/CVPRW50498.2020.00203.

[37] Sinong Wang et al. "Linformer: Self-attention with linear complexity". In: arXiv preprint arXiv:2006.04768 (2020).

[38] Mehta S., Rastegari M., "MobileViT: Light-weight, General-purpose, and Mobile-friendly Vision Transformer". International Conference in Learning Representation. 2022. Available at:http://arxiv.org/abs/2110.0 2178

[39] Chollet F. "Xception: Deep Learning with Depthwise Separable Convolutions," 2017 IEEE Conference on Computer Vision and Pattern Recognition (CVPR), Honolulu, HI, USA, 2017, pp. 1800-1807, doi: 10.1109/CVPR.2017.195.

[40] Lin J. et al. "MCUNet: Tiny Deep Learning on IoT Devices". Annual Conference on Neural Information Processing Systems (NeurIPS 2020) Nov. 2020. doi: 10.48550/arXiv.2007.10319.

[41] Lin J. et al., "MCUNetV2: Memory-Efficient Patch-based Inference for Tiny Deep Learning". Oct. 2021. arXiv preprint. doi: 10.48550/arXiv.2110.15352.

[42] C. Szegedy et al., "Going deeper with convolutions," 2015 IEEE Conference on Computer Vision and Pattern Recognition (CVPR), Boston, MA, USA, 2015, pp. 1-9, doi: 10.1109/CVPR.2015.7298594.

[43] Szegedy C., Vanhoucke V., Ioffe S., Shlens J., Wojna Z., "Rethinking the Inception Architecture for Computer Vision," 2016 IEEE Conference on Computer Vision and Pattern Recognition (CVPR), Las Vegas, NV, USA, 2016, pp. 2818-2826, doi: 10.1109/CVPR.2016.308.

[44] Ioffe S., Szegedy C. "Batch Normalization: Accelerating Deep Network Training by Reducing Internal Covariate Shift". Proceedings of the 32nd International Conference on Machine Learning (ICML) 37:448-456, 2015.Mar. 2015. doi: 10.48550/arXiv.1502.03167.

[45] Huang G., Liu Z., Van Der Maaten L. Weinberger K. Q., "Densely Connected Convolutional Networks," 2017 IEEE Conference on Computer Vision and Pattern Recognition (CVPR), Honolulu, HI, USA, 2017, pp. 2261-2269, doi: 10.1109/CVPR.2017.243.

[46] Xiaohan Ding et al. "Diverse Branch Block: Building a Convolution as an Inception-like Unit". In: 2021 IEEE/CVF Conference on Computer Vision and Pattern Recognition (CVPR). Nashville, TN, USA: IEEE, June 2021, pp. 10881–10890. doi: 10.1109/CVPR46437.2021. 01074.

[47] Vasu P, Gabriel J., Zhu J., Tuzel O., Ranjan A., "MobileOne: An Improved One millisecond Mobile Backbone". Mar. 2023. arXiv preprint, url:http://arxiv.org/abs/2206.04040

[48] Carion N., Massa F., Synnaeve G., Usunier N., Kirillov A., Zagoruyko S., "End-to-End Object Detection with Transformers". In: ECCV 2020. LNCS vol 12346. doi: 10.1007/978-3-030-58452-8_13.

[49] Adekar S.N., Chaurasia A. "MobileViTv3: Mobile-Friendly Vision Transformer with Simple and Effective Fusion of Local, Global and Input Features". Oct. 2022. arXiv preprint. doi: 10.48550/arXiv.2209.15159.

[50] Mehta S, Rastegari M. "Separable Self-attention for Mobile Vision Transformers". June 202. arXiv preprint. doi: 10.48550/arXiv.2206.02680.

[51] Liu Z. et al., "Swin Transformer: Hierarchical Vision Transformer using Shifted Windows," 2021 IEEE/CVF International Conference on Computer Vision (ICCV), Montreal, QC, Canada, 2021, pp. 9992-10002, doi: 10.1109/ICCV48922.2021.00986.

[52] Dosovitskiy A. et al. "An Image is Worth 16x16 Words: Transformers for Image Recognition at Scale", 9th International Conference on Learning Representations (ICLR 2021). May 2021. doi: 10.48550/arXiv.2010.11929.

[53] Xie S., Girshick R., Dollár P., Tu Z., He K., "Aggregated Residual Transformations for Deep Neural Networks," 2017 IEEE Conference on Computer Vision and Pattern Recognition (CVPR), Honolulu, HI, USA, 2017, pp. 5987-5995. Available at:https://doi.org/10.1109/CVPR.2017. 634. Deng J., Dong W., Socher R., et al., "ImageNet: A Large-Scale Hierarchical Image Database". 2009 IEEE Conference on Computer Vision and Pattern Recognition, Miami, 20-25 June 2009, 248-255. doi: 10.1109/CVPR.2009.5206848.

[54] Barret Zoph et al. "Learning Transferable Architectures for Scalable Image Recognition". In: 2018 IEEE/CVF Conference on Computer Vision and Pattern Recognition. Salt Lake City, UT: IEEE, June 2018, pp. 8697–8710. doi: 10.1109/CVPR.2018.00907.

[55] Liu C. et al. "Progressive Neural Architecture Search". In: ECCV 2018. LNCS vol 11205. doi: 10.1007/978-3-030-01246-5_2

[56] Mingxing T. et al. "MnasNet: Platform-Aware Neural Architecture Search for Mobile". In: 2019 IEEE/CVF Conference on Computer Vision and Pattern Recognition (CVPR). Long Beach, CA, USA: IEEE, June 2019, pp. 2815–2823. doi: 10.1109/CVPR.2019.00293.

6

Natural Language Conditioned Planning of Complex Robotics Tasks

**Toms Eduards Zinars, Oskars Vismanis, Peteris Racinskis,
Janis Arents, and Modris Greitans**

Institute of Electronics and Computer Science, Latvia

Abstract

As natural language processing advances in the field of robotics, enabling
seamless human-robot interaction, it becomes imperative to identify the most
effective approach for conditioning complex robotics tasks using natural
language commands. This article reviews various state-of-the-art methods
for natural language-conditioned planning, with a particular focus on mobile
manipulation. The authors explore and review different architectures and
techniques to comprehend, interpret, and execute natural language commands. Challenges are identified along the way, and conceptual architecture
is proposed to tackle them in an efficient manner.

Keywords: natural language processing, mobile manipulation, action primitives, edge AI.

6.1 Introduction

Everyday interactions between people are usually performed in a very casual
manner through natural language, or NL, as it is an comfortable way of
communication. This has been carried over to our appliances, such as phones
and cars, with the use of voice commands. It logically follows that robotic
assistants, be they industrial or service, which operate in an environment with
people of various machinery handling skill levels, would benefit from a user

DOI: 10.1201/9788770041027-6   131

interface utilizing Natural Language Processing (NLP) techniques to process unstructured input. As the environments in which these systems get deployed are diverse and varied, a general understanding of language and its relation to objects is crucial to a rational implementation.

Large Language Models (LLM) [1][2] have shown to be very capable of tackling the task of inferring NL inputs for commanding a robotic agent. They can be made to operate with multi-modal information [3], but they often rely on cloud service models that require immense computing resource [4][5][6]. The possibility exists to use smaller models that can be run locally [7][2]. Robotic systems are typically specialized to operate under certain conditions, and a trained technician must perform any new adaptations. However, as real-life environments are usually unique in their layout and the objects they contain, it can be hard to predict what the robotic agent needs to know. In section 1.2, we do a quick overview of NL processing, LLM and multi-modal embeddings and their recent implementations and uses in planning for robotic systems.

To function in such environments, a more general approach for representing robot tasks can be used, as, fundamentally, a robot system can be thought of as a handful of basic operations, but they are usually very task-dependant. One such approach is using action primitives. They provide the high-level planner with an abstract, symbolic representation of available actions so a task plan can be made. Another role they fulfill is low-level planning, a set of functions that can be adapted to specific scenarios and reused for different tasks [8]. Details about action primitives, their synthesis and implementations are described in section 1.3.

The primary type of robot control we address with the approach outlined in this chapter is mobile manipulation, which can be described as the joint control of a mobile base and a manipulator arm. The overall workflow of such a system can be generalized to receiving, planning and executing a task. In our conceptual architecture, the function of receiving a task is done with NL commands. For this, a two-stage technique is proposed. A high-level LLM-based algorithm for mapping NL commands into a constrained space of action primitives, and a library of action primitives at the low level, which is further elaborated on in section 1.5. This chapter also includes section 1.4, where we provide a brief overview of identified challenges and issues regarding both the practical use of NLP and the implementation of Action Primitives, while section 1.6 concludes the chapter with our views on future developments.

6.2 Natural Language Processing for Robotics

Natural language is vast and vague, so much so that people often have problems understanding each other. An NL command can consist of a request that only makes sense in the context of the situation and the environment. It is not enough for a robotic system to just deduce what is being asked of it but also to be able to ground this information in the environment it finds itself in and act upon through a concise and safe plan of actions.

The current state-of-the-art performance in NLP can be found in language models based on transformer architecture [9], specifically the LLMs. Starting at 10 billion, typically having 100-300 and with a couple exceeding 1 trillion, these LLMs have demonstrated impressive performance across various language-related tasks [1][10].

6.2.1 Large language models

Large Language Models are powerful natural language interpreters [11]. An important quality they share with the smaller models is multitask learning – the ability for one model to become relatively proficient at several different tasks but masters of none [12]. However, as their size grows and performance increases [13], these models begin to overtake previous-generation specialist fine tuned models [10][11].

Training large language models from scratch is a very expensive process in terms of energy, computing and time, typically requiring massive clusters of dedicated high-end hardware that train them non-stop for several weeks [2]. As this is only feasible for large tech corporations such as Google, Meta, OpenAI, Amazon and Huawei [2], a pre-trained model can be specialized for a downstream task through fine-tuning - using a specialized dataset to introduce into the model specific knowledge or teach it new operations all together [14], making them more accessible for specific tasks.

When the parameter count reaches into the tens and hundreds of billions, the language models begin to exhibit new qualities that are not present in their smaller counterparts, referred to as emergent abilities [1][15], three prominent ones, as highlighted in [1], are:

- In-context learning [11]- the ability to perform tasks not part of the training corpus, based on an instruction and several input-output examples(few-shot) provided in the prompt;
- The instruction following [14] relies on fine-tuning a model using a dataset containing natural language instructions, which results in

improved zero-shot (no example given) prompting performance for unseen tasks;

• Step-by-step [16] improves the model's complex reasoning by leveraging chain of thought prompting by adding step-by-step instructions to the examples for a few-shot task prompt.

As fine-tuning takes computing resources and time, many pre-trained models are fully capable of being used out of the box [11]. By carefully structuring the input prompts, it is often possible to condition the model to provide the desired output for the task at hand [17], an approach known as prompt engineering [1].

As a way to improve the fine-tuning process, a lot of work has been done in developing various parameter-efficient fine-tuning methods (PEFTs) [18] that consider fine-tuning only parts of the overall model. Instead of having several different fine-tuned copies of the same LLM weights, using methods such as Low-Rank Adapters (LoRA) [19] one can have a single instance of an LLM and then simply apply the corresponding fine-tuned adaptation, saving on space and compute resource.

Running the LLM inference process, even on the smaller models, requires capable hardware, primarily GPUs with sufficient VRAM [7]. The model's parameters are typically stored in a 16-bit float format (FP16), translating to roughly 2GBs for every 1 billion parameters. A remedy for this issue is model quantization, a method where the parameters of the model are converted into smaller 8-, 4-, 3- and even 2-bit formats [20], which (plural) can reduce the required VRAM down from 14GB to roughly 4 GBs for a 7B model (7B representing 7 billion parameters) when using 4-bit quantization, with marginal loss to performance [7]. The requirements can be further reduced by using methods that share inference between CPU and GPUs [21], which provides perspectives for application in edge AI.

As language models have been trained on general data such as textual information sourced from books or the internet and/or coding languages [1][2][22], they gain a broad internal knowledge base that can be leveraged to create a human-machine interface capable of decoding obscure NL requests into actions understandable to a robotic agent system [4][17].

6.2.2 Multi-modal embeddings

The recent progress in autoregressive and sequence-to-sequence NLP processing with LLMs has enabled a number of related advances. In particular, the abstract vector nature of the tokens being processed by transformers has been exploited to create mappings between radically different data modalities. CLIP [3], short for Contrastive Language-Image Pre-training, is a notable example. It jointly trains an image classifier and text encoder on image-caption pairs. Each model outputs a vector in the same latent space. The cosine similarity of vectors corresponding to matching image-caption pairs is maximized, while that between all others is minimized. The result is a pair of models capable of mapping the greatly dissimilar image and text input spaces to a common latent "concept" space.

CLIP and similar systems have since been commonly described as vision-language models (VLMs). Subsequent work, such as *LSeg* [23] and *ConceptFusion* [24], has been done to extend the vision model in a VLM to produce segmentation maps – embeddings for each image in a pixel. These have subsequently found use in robotics, particularly in creating maps amenable to natural language queries. For example, in [25], a 2-dimensional grid map is constructed using LSeg and depth imagery, which can then be used to find navigation goals using text prompts. ConceptFusion [24] expands upon the mapping problem, producing 3-dimensional embedding-tagged point clouds. Some approaches do away with explicit maps entirely, instead using a Neural Radiance Field to predict the embedding associated with any point in the environment directly [26].

The ability of transformer models to map between and autoregressively generate sequences of arbitrary vectors has been directly exploited for robot control in works such as [27], where robot actions are predicted directly from text prompts and images of the scene in which the robot should operate. The inputs need not be limited to a single type of embedding — in [28] and [29], a large transformer is trained to operate on input sequences containing multiple types of embeddings — such as VLM tokens, robot state encodings, scene representations and past actions — with PaLM-E in [29] being directly based on a pre-trained LLM.

6.2.3 Recent implementations of high-level planning for mobile manipulation

The practical implementation of language models for use in high-level planning of mobile manipulator systems has taken various approaches. Some approaches use the language models to extract language features that are further passed into more specialized modules for processing [30][31], others use the language models as active elements of the planning process [5][6][32], and others yet use the models for low-level planning [33]. Some models perform their own mapping of the environment through computer vision [30][34], but it would seem that map integration is an underutilized solution, though some works are exploring combining embeddings from the language model with embeddings stored in a semantic map [6].

Many of the highlighted works rely on prompting and using pre-trained models [4][5][6][32][33][35]. The importance of proper prompting technique is explored in [19], which presents a method for selecting and formatting prompts to elicit outputs usable in robotic systems. They define a starting prompt that describes the role the LLM is supposed to play and condition it to respond only when directly prompted to by a specific keyword. That is followed up with a sequence of instructions, explanations and templates that describe the desired output format and contents. The prompt is finished up by providing several examples of how the output should look. The ability to provide NL feedback to improve and correct mistakes during inference is also showcased.

While not an example of a natural language command, ProgPrompt [32] takes an input prompt of Python code containing imports of action primitives, a list of available objects, example tasks and the start of the desired operation. The LLM then returns a generated plan in Python that uses assertions to ensure a feedback loop once the agent encounters variables in the environment and can successfully execute the appropriate action.

Lang2LTL [6] utilizes a modular system where LLMs are used to perform several subtasks in the interest of generating a grounded relation to objects and places that the system stores in a database. One module is tasked with extracting place names from the request prompt. These extracted names are then compared to the database objects through embedding cosine similarity, and then generalizing the input request with substitutions and passing it through a fine-tuned LLM symbolic translator that generates the LTL formula, finished by inserting the found database objects in their respective substitution locations.

Text2Motion [5] utilizes an LLM model that performs task planning in conjunction with geometric feasibility planning that evaluates if the plan generated by the LLM is valid or not. They evaluated planning the whole sequence of actions and then validating it, planning and validating each individual step of the sequence and a hybrid system that tries creating a full plan, falls back to individual step in case of a failure, then tries finishing the plan fully again. For implementation, they rely on OpenAI's GPT series [36] and execute their system through prompt engineering.

Language to Rewards [33] employs a two-stage LLM setup in the form of a Reward Translator, where the first LLM (Motion descriptor) is used to translate the input sentence into a structured natural language instruction. The second LLM (Reward Coder) then generates a usable code in the form of reward functions that can be passed directly to a low-level motion controller, skipping the use of action primitives altogether. Both LLMs are conditioned by leveraging in-context learning by prompts. The first one contains templates to use when creating the task description, while the second is prompted with a general program description. Both prompts contain a list of instructions to guide inference to the desired result. While this approach doesn't perform long-horizon tasks, it does showcase the possibility of using LLMs for low-level planning to some degree.

Say-Can [3] explored the issue of grounding an LLM planning system in the real world, as without any feedback elements or information about the current environment, the language model can propose logical but contextually impossible solutions such as suggesting using a vacuum cleaner when one isn't available. They achieved this by using a two-part system - the LLM provides probabilities for action relevance to the given task, while a *value function* provides probabilities of how likely it is to succeed in doing specific actions. The multiplication of these two values is chosen as the action for the plan to perform.

Inner Monologue [35] explores using feedback mechanisms to improve task completion. By being able to receive information from the environment in the form of language input, such as sensory data about detected objects or whether the planned action was successful or not, the agent can attempt to perform the action again or replan, whereas without such feedback the agent would fail the task outright. When the agent is met with an ambiguous situation, such as a request for "a drink", by asking the user for clarification, it can form a dialogue that helps execute the task more successfully. They also question if the answering could be done by another LLM as well.

There are doubts by some if LLMs are reliable enough to be used for planning operations [31][37] but do recognize their utility as language interpreters. One such implementation is proposed in [31] with LLM+P, a language model coupled with a classical planner. A classical planner provides proven ability in task solving, while the LLM can provide its understanding of language to be able to interpret a large amount of tasks and then translate them to a structured planning language such as PDDL [47].

Language is only one part of a system meant to operate in an environment. The ALFRED benchmark was introduced in 2020 as a way to test agent systems that use both natural language instructions and ego-centric vision [38]. While not the first, it did combine several functions to create a benchmark that tests proposed systems in a non-reversible, partially observable environment. Models are tasked with solving basic household tasks within a limited number of actions, which typically involve moving and modifying objects using tools or special locations, requiring a specific sequence of actions to execute. Many models also build a map representation of the environment [30][34]. The baseline model relied on an LSTM (long-short-term memory, predecessor to the transformer architecture) based language model that managed to achieve only a 0.4% success rate in the unseen tests [38]. Later attempts would implement PLMs such as BERT [39] and improve the success rate to 50% [34]. At the time of writing, the best-performing models that have available materials are Prompter [30] and CAPEAM [34], both utilizing BERT for their language processing. BERT is an older language model (from late 2018) with sub-1 billion parameters, far from state-of-the-art in language models, making a direct comparison hard as LLM-based systems seem to rely on their own evaluation methods [3].

Prompter [30] utilizes its language model in a semantic search module, using natural relations between objects (apples found in kitchens, toothbrushes in bathrooms), speeding up the search. The benefits of using a language model for such a role is leveraging its inherent knowledge of language to determine word relations, whereas previous methods relied on using additional training. For its vision substream, based on [39], the agent creates a 2D top-down semantic map from images it received through its ego-centric vision, which is a 2D RGB image that is processed into a depth map and segmented to create masks.

CAPEAM [34] uses a fine-tuned BERT implementation for predicting what predefined role fits what words from the input sentence. Context Aware Planning module uses a sub-goal planning element that first finds a general template to use based on input requests and then a so-called "detailed planner" to insert the contextual information into predefined places. Environment Aware Memory is responsible for vision operations, utilizing memorization for object locations as well as saving previous segmentation masks, as it was found it helps to identify objects that become obscured through later actions.

6.3 Action Primitives for Mobile Manipulation

In [40], the authors define the concept of *situated robotics*, which describes robotic systems in complex, dynamic or, in other words, *situated* environments. The amount of an environment's situatedness directly affects the complexity of the robot control system and its need to adapt to new situations, which shows that the more complex the environment, the more complex the control system needs to be. Derived from different action definitions, different approaches to robot control exist, such as reactive control, deliberative control, hybrid control (a combination of the first two), and behaviour-based control [40]. Reactive Controls can be referred to as an IF...THEN rule interpreter while Deliberative control - as functioning in a higher level. Behaviour-based control, on the other hand, functions a bit differently; at its core is the concept of behaviours, which are functionalities varying in complexity that get activated depending on their predefined inputs, which can be sensory data or other behaviours, and they output control commands for actuators or other behaviours. Where the other control type's lack in either computation efficiency or complexity, the behaviour-based control can manage a combination of the two, that is easier to engineer and upgrade than hybrid control [40]. From the concept of behaviours, abstract or primitive behaviour can be derived, which is a more general function made to be reusable in different scenarios. These are often referred to in the literature as the **action primitives** (aka manipulation primitives, task primitives, skills etc.).

To plan and execute tasks in situated environments, some form of Task and Motion Planning (TAMP) is usually required, as seen in [4][41][42][43], and for that, it is best to have representations of the environment and available robot actions. As mentioned before, dynamic environments can have many different actions fulfilled in them and engineering all of them can be a time-consuming process [if even possible]. A better approach might be to use said **action primitives**, which would be task-specific only in the execution phase, depending on sensory inputs. This representation of behaviours allows for a more general form of activation conditions as these are usually atomic functions that do only one thing, but not in a way feedback control would be managed (for example, move by a certain angle) [8][44][40]. The granularity of primitives depends on the usage, but in robotics, it is usually a control command to make a robot move. As action primitives are usually computationally light, one system can be made to work for different tasks.

Primitives can be divided by their usage, the simplest form being the primitive itself. After primitives come actions, and after that - activities [44].

Taking as an example a robot doing picking, that could be considered an action, which would be made up of multiple primitives, in this case, moving to the object and closing the gripper. The case of putting multiple actions in a sequence, such as picking and placing, could be referred to as an activity of moving an object. This way of referencing actions is especially useful for task planning, as the symbolic representation of a task does not always include all the steps needed to complete it [41]. If, for example, the task of picking up an object and placing it somewhere is given, the task planner does not need to think about the specific primitives needed, such as moving the arm to the specific spot and closing the gripper, it just needs to make a sequence of actions, that fulfil the task. The primitives are then left for the motion planner to check geometrically if the task plan is feasible. Many systems would then use a so-called action library or a set of skills [41][42][45] that can be used by the task planner to know what the system is capable of and use it when making plans.

Symbolically, actions and the action-state relations can be predefined using task planning languages such as STRIPS [46] or its successor, PDDL [47], or it can be done during the planning process using Large Language Model (LLM) prompt engineering [48]. Regarding the environment, the representation can be about locations, objects, etc. Actions refer to what tools the robot has at its disposal, in other words, what it can do to accomplish its tasks.

6.3.1 Methods for creating primitives

There can be many kinds of primitives made for specific applications, which means that creating them is a process of its own, and there are different ways of synthesis. For example, Jeon et al. [41] create a service robot application, which utilizes an action library in which one of the actions is *hold_object*. This action can be decomposed into two action primitives – *approach_object* and *close_hand*. For task planning purposes, actions, action primitives and their interrelations are represented using the PDDL language. This allows them to be used with PDDL-supported planners. In addition, it predefines the action's preconditions, effects, requirements, etc. This method deeply relies on the engineer's capabilities and understanding of the tasks they're making the primitives for. Also, in the case of wanting to add new functionality, it can be a long process, depending on the complexity of the task.

Action primitives can also be extracted from human motion via imitation learning (IL) [49]. In [50], manually segmented human motion capture data is used with a spatio-temporal non-linear dimension reduction technique to cluster similar segments of motion into generalized primitives. Similarly, in [43], imitation learning is used, in terms of behavioural cloning (BC), to learn action primitives, but their combinations into actions are then learned with reinforcement learning (RL). With these methods, the system is able to do the task of pouring cereal into a bowl. In [51], however, the stereotypical motions of a human picking up a cup are recorded and used as a basis for actions, though in the form of dynamic movement primitives (DMP).

A different approach to creating primitives is only making them when needed. Gizzi et al. [42] look at using action primitives as a way of creative problem-solving. They use the definition of a MacGyver problem [52], which describes an environment that has everything necessary for successful task execution, and the robot has all the tools it needs. Still, the specific approach it must take is unknown. The system begins with a set of predefined actions and is tasked to do an indirect task, such as reaching an obstructed object or location. Whenever met with a situation where the robot cannot fulfil its task with the actions it knows, it starts generating combinations of available objects and interactions with them using the available actions. The environment and actions are described using PDDL. In this case, the actions available are *obtain_object* and *press_button*, and the robot is tasked with reaching an object on the other side of a wall. There are also buttons present. When the initial *obtain_object* fails, it starts generating different combinations of actions and tries executing the feasible ones one by one until it manages to hold down one of the buttons to move the wall.

In the case of service robots or any robots that might work in environments simultaneously with humans, there is also a positive effect when creating the primitives to function similarly to human motion, as humans can better understand them from the point of predictability. This approach also allows for easier training using imitation learning [51].

6.3.2 Action primitive implementations

In relation to mobile manipulation or manipulation in general, the action primitive has been used consistently in systems that do more general or environment-adaptive tasks. In [42], action primitives are used as a basis for finding solutions to given problems in the sense of finding new primitives that would be applicable to the situation.

In [53], such action primitives as *pulling*, *pushing*, *grasping* and *pivoting* are included. A dual-arm robot ABB YuMi with custom tactile sensors is used. With the predefined action library and tactile feedback combination, the system is able to do dexterous manipulation based on robot/object interaction plans. This same setup is used in [54] to manipulate rigid objects based on pointcloud data using long-horizon planning. The planning problem is defined as an action primitive sequencing, where the symbolic representation of actions as action primitives allows for the planner to set aside the reasoning about robot-object dynamics.

For service robots, action primitives are used to create and execute plans for such tasks as pouring cereal [43] or juice [41]. In [43], basic action primitives are learned and then combined into such actions as *pick*, *bowl* (picking an object and placing it in a bowl) and *breakfast* (moving objects in a pouring manner directed toward the bowl). For similar tasks, [41] uses such primitives as *approach_object, close_hand, and move_arm*. These are then used in plans to accomplish tasks like moving an object out of the way and then moving a package in a pouring motion.

6.4 Identified Challenges

Robotic agents designed to work alongside people are under great scrutiny, as such systems must be, first and foremost, safe, adequately efficient and easy to use. Such systems have to be robust, with no room for ambiguity. Yet, the datasets that language models are trained on come with biases that the model inherits and these biases can affect the inference process and result in seemingly random erroneous outputs [1][37] or repetitive cycles [5]. There is also a general risk of hallucination from LLM providing absurd plans to the robotic agent system or failing to generate anything at all.

Many promising implementations rely on using the closed-source GPT series for their research [17][5][6][33]. While these models are state-of-the-art and the main target for other models to beat [22], a real implementation for any mobile agent lacks practical autonomy if it must have a constant connection to a cloud service. If it is true that certain traits of language models are limited to the larger models and begin appearing only at 100B+ parameters [1][15], even with quantization, these larger models would require powerful GPUs that are impractical to be placed onto a physical agent to ensure autonomous operation, limiting the system to require a local network where a remote resource provides inference to the system. This impedes taking advantage of the LLMs emergent abilities for edge AI applications.

One of the main focuses of NL commanded mobile manipulation systems is higher autonomy, but they often still lack in their abilities. For example, a clear bottleneck of such systems is the range and capabilities of the skills they possess as [4], even if the NL command can be translated to the system correctly, it cannot execute something it does not know how.

Another challenge, directly involving manipulation, can be addressed in the form of object recognition and grasping. A real environment usually has many different types of objects, and a robot operating in such an environment would be expected to be able to manipulate any of them. Modern approaches manage such a problem using 3D model databases from the internet [51] or pre-trained vision-language models [55] that can recognize previously unseen objects. Grasping them is then dependent on grasp point recognition [56][57] and the quality of active primitives used.

6.5 Conceptual Architecture

Based on what has been covered, we have considered various implementations for robotics systems with natural language commands. There is no one definitive architecture (as can be seen by examples in chapter 6.2.3), given differing sizes, scopes and environments in which such agents are expected to operate. The underlying principles we believe such a system would need to have are the need to process the language input, a way to locate itself and objects in the environment, plan its actions and finally execute the plans at the physical level, with a feedback system to account for changing factors in the environment or the user's request.

In Figure 6.1 we present a simplified interpretation of how such a system could be structured in two distinct main modules or levels – the language processing module and the execution module. Other works have also approached the problem with some type of two-part design [4][34]. What may seem missing from this schematic is a dedicated planning module, but here it is understood that high-level or task planning is done by the NLP module, whereas low-level planning or motion planning is done in the execution module.

At the core of the NLP module would be an LLM that could be run locally, at least on a cluster, but which one fits this use case needs to be investigated. The module's task is to interpret the user's input request in a way that the rest of the system may act upon it. For successful communication between the levels, the execution module provides the NLP module with a description of what it can do and how the language module output needs to be formatted, allowing for new capabilities to be introduced to and utilized by the system

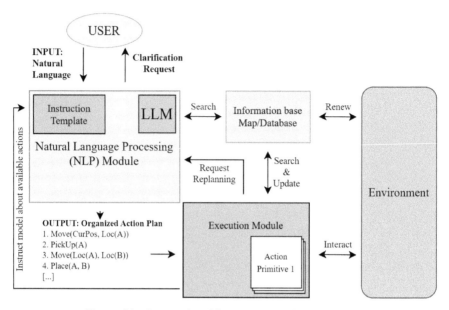

Figure 6.1   Proposed Mobile Manipulator Control System

that were not originally accounted for. An interface with a map (or database) grounds the language module in its current environment and permits the NLP module access to task-relevant information. It would be preferable as well for the NLP module to be able to take received feedback from the execution module and present it to the user for clarification.

The execution module consists of an action library, which is a set of action primitives the agent can execute to accomplish tasks. The module is responsible for physical execution, so motion planning is also done here. Whenever a task plan is received, this module checks its geometrical feasibility and, in case of a failure, requests for a replanning. Once a feasible plan is made, the module executes the task sequence.

The main benefits of such a layout is the idea that the information about the environment is contained within the semantic map, while the LLM possesses general linguistic knowledge. As action primitives should be the same regardless of where it is, by combining information extracted from the input with information available on the map, the system is not hard-coded to a particular place. By relying on a smaller and locally run LLM instance, the hope is to ensure the ability of the system to operate successfully in an edge AI use case.

6.6 Conclusions and Outlook

Natural language application in robotics is an ever more relevant field of research and development. The rise of LLMs has made applying general language understanding to computer systems seem deceptively trivial, but there is still much ambiguity to overcome. We hope to see the field develop in both directions - more research done on larger models to see how much they are capable of, as well as more development to bring these high-level abilities down to smaller and smaller model sizes to enable true edge AI applications.

When working in complex environments, action primitives can be used as a powerful tool to generalize actions available to a robotic system. This can be useful both with task and motion planning as they allow for these two processes to be less intertwined without affecting their effectiveness. There are different approaches to creating action primitives, and the future seems to be headed towards automated synthesis with different machine-learning techniques. This chapter proposes a conceptual architecture for NL-commanded mobile manipulation, consisting of an NLP module for command interpreting and high-level planning and an execution module that utilizes action primitives for low-level planning and execution.

Acknowledgements

This research was conducted as part of the EdgeAI "Edge AI Technologies for Optimised Performance Embedded Processing" project, which has received funding from KDT JU under grant agreement No 101097300. The KDT JU receives support from the European Union's Horizon Europe research and innovation program and Austria, Belgium, France, Greece, Italy, Latvia, Luxembourg, Netherlands, and Norway.

References

[1] Zhao, W. X., Zhou, K., Li, J., Tang, T., Wang, X., Hou, Y., Min, Y., Zhang, B., Zhang, J., Dong, Z., Du, Y., Yang, C., Chen, Y., Chen, Z., Jiang, J., Ren, R., Li, Y., Tang, X., Liu, Z., Liu, P., Nie, J., & Wen, J. (2023). A Survey of Large Language Models. *ArXiv, abs/2303.18223.*

[2] Naveed, H., Khan, A. U., Qiu, S., Saqib, M., Anwar, S., Usman, M., Barnes, N., & Mian, A. S. (2023). A Comprehensive Overview of Large Language Models. *ArXiv, abs/2307.06435.*

[3] Radford, Alec, Jong Wook Kim, Chris Hallacy, Aditya Ramesh, Gabriel Goh, Sandhini Agarwal, Girish Sastry, Amanda Askell, Pamela Mishkin, Jack Clark, Gretchen Krueger and Ilya Sutskever. "Learning Transferable Visual Models From Natural Language Supervision." International Conference on Machine Learning (2021).

[4] Ahn M., Brohan A., Brown N., Chebotar Y., Cortes O., David B., Finn C., Gopalakrishnan K., Hausman K., Herzog A., Ho D., Hsu J., Ibarz J., Ichter B., Irpan A., Jang E., Ruano R. J., Jeffrey K., Jesmonth S., Joshi N. J., Julian R. C., Kalashnikov D., Kuang Y., Lee K.-H., Levine S., Lu Y., Luu L., Parada C., Pastor P., Quiambao J., Rao K., Rettinghouse J., Reyes D. M., Sermanet P., Sievers N., Tan C., Toshev A., Vanhoucke V., Xia F., Xiao T., Xu P., Xu S., Yan M. Do As I Can, Not As I Say: Grounding Language in Robotic Affordances. Conference on Robot Learning, 2022.

[5] Lin, K., Agia, C., Migimatsu, T., Pavone, M., & Bohg, J. (2023). Text2Motion: From Natural Language Instructions to Feasible Plans. *ArXiv, abs/2303.12153*.

[6] Liu, J., Yang, Z., Idrees, I., Liang, S., Schornstein, B., Tellex, S., & Shah, A. (2023). Lang2LTL: Translating Natural Language Commands to Temporal Robot Task Specification. *ArXiv, abs/2302.11649*.

[7] Frantar, E., Ashkboos, S., Hoefler, T., & Alistarh, D. (2022). GPTQ: Accurate Post-Training Quantization for Generative Pre-trained Transformers. *ArXiv, abs/2210.17323*.

[8] Schaal S., Ijspeert A., Billard A. Computational approaches to motor learning by imitation. In: Philosophical Transactions of the Royal Society B: Biological Sciences, 2003, 358(1431), 537–547.

[9] Vaswani, A., Shazeer, N.M., Parmar, N., Uszkoreit, J., Jones, L., Gomez, A.N., Kaiser, L., & Polosukhin, I. (2017). Attention is All you Need. NIPS.

[10] OpenAI (2023). GPT-4 Technical Report. *ArXiv, abs/2303.08774*.

[11] Brown, T.B., Mann, B., Ryder, N., Subbiah, M., Kaplan, J., Dhariwal, P., Neelakantan, A., Shyam, P., Sastry, G., Askell, A., Agarwal, S., Herbert-Voss, A., Krueger, G., Henighan, T. J., Child, R., Ramesh, A., Ziegler, D.M., Wu, J., Winter, C., Hesse, C., Chen, M., Sigler, E., Litwin, M., Gray, S., Chess, B., Clark, J., Berner, C., McCandlish, S., Radford, A., Sutskever, I., & Amodei, D. (2020). Language Models are Few-Shot Learners. ArXiv, abs/2005.14165.

[12] Radford, A., Wu, J., Child, R., Luan, D., Amodei, D., & Sutskever, I. (2019). Language Models are Unsupervised Multitask Learners.

[13] Kaplan, J., McCHuang, W., Xia, F., Xiao, T., Chan, H., Liang, J., Florence, P.R., Zeng, A., Tompson, J., Mordatch, I., Chebotar, Y., Sermanet, P., Brown, N., Jackson, T., Luu, L., Levine, S., Hausman, K., & Ichter, B. (2022). Inner Monologue: Embodied Reasoning through Planning with Language Models. Conference on Robot Learning.andlish, S., Henighan, T. J., Brown, T. B., Chess, B., Child, R., Gray, S., Radford, A., Wu, J., & Amodei, D. (2020). Scaling Laws for Neural Language Models. ArXiv, abs/2001.08361.

[14] Wei, J., Bosma, M., Zhao, V., Guu, K., Yu, A. W., Lester, B., Du, N., Dai, A. M., & Le, Q. V. (2021). Finetuned Language Models Are Zero-Shot Learners. ArXiv, abs/2109.01652.

[15] Wei, J., Tay, Y., Bommasani, R., Raffel, C., Zoph, B., Borgeaud, S., Yogatama, D., Bosma, M., Zhou, D., Metzler, D., Chi, E. H., Hashimoto, T., Vinyals, O., Liang, P., Dean, J., & Fedus, W. (2022). Emergent Abilities of Large Language Models. Trans. Mach. Learn. Res., 2022.

[16] Wei, J., Wang, X., Schuurmans, D., Bosma, M., Chi, E. H., Xia, F., Le, Q., & Zhou, D. (2022). Chain of Thought Prompting Elicits Reasoning in Large Language Models. ArXiv, abs/2201.11903.

[17] Wake, N., Kanehira, A., Sasabuchi, K., Takamatsu, J., & Ikeuchi, K. (2023). ChatGPT Empowered Long-Step Robot Control in Various Environments: A Case Application. ArXiv, abs/2304.03893.

[18] Lialin, V., Deshpande, V., & Rumshisky, A. (2023). Scaling Down to Scale Up: A Guide to Parameter-Efficient Fine-Tuning. ArXiv, abs/2303.15647.

[19] Hu, J. E., Shen, Y., Wallis, P., Allen-Zhu, Z., Li, Y., Wang, S., & Chen, W. (2021). LoRA: Low-Rank Adaptation of Large Language Models. ArXiv, abs/2106.09685.

[20] Chee, J., Cai, Y., Kuleshov, V., & Sa, C.D. (2023). QuIP: 2-Bit Quantization of Large Language Models With Guarantees. ArXiv, abs/2307.13304.

[21] Gerganov, G. Port of Facebook's LLaMA model in C/C++. Available at: https://github.com/ggerganov/llama.cpp [Accessed August 23, 2023]

[22] Rozière, B., Gehring, J., Gloeckle, F., Sootla, S., Gat, I., Tan, X. E., Adi, Y., Liu, J., Remez, T., Rapin, J., Kozhevnikov, A., Evtimov, I., Bitton, J., Bhatt, M. P., Ferrer, C.C., Grattafiori, A., Xiong, W., D'efossez, A., Copet, J., Azhar, F., Touvron, H., Martin, L., Usunier, N., Scialom, T., & Synnaeve, G. (2023). Code Llama: Open Foundation Models for Code.

[23] Li, Boyi, Kilian Q. Weinberger, Serge J. Belongie, Vladlen Koltun and René Ranftl. "Language-driven Semantic Segmentation." ArXiv abs/2201.03546 (2022).

[24] Jatavallabhula, Krishna Murthy, Ali Kuwajerwala, Qiao Gu, Mohd. Omama, Tao Chen, Shuang Li, Ganesh Iyer, Soroush Saryazdi, Nikhil Varma Keetha, Ayush Kumar Tewari, Joshua B. Tenenbaum, Celso M. de Melo, M. Krishna, Liam Paull, Florian Shkurti and Antonio Torralba. "ConceptFusion: Open-set Multi-modal 3D Mapping." ArXiv abs/2302.07241 (2023).

[25] Huang, Chen, Oier Mees, Andy Zeng and Wolfram Burgard. "Visual Language Maps for Robot Navigation." 2023 IEEE International Conference on Robotics and Automation (ICRA) (2022): 10608-10615.

[26] Shafiullah, Nur Muhammad (Mahi), Chris Paxton, Lerrel Pinto, Soumith Chintala and Arthur Szlam. "CLIP-Fields: Weakly Supervised Semantic Fields for Robotic Memory." ArXiv abs/2210.05663 (2022).

[27] Brohan, Anthony, Noah Brown, Justice Carbajal, Yevgen Chebotar, Krzysztof Choromanski, Tianli Ding, Danny Driess, Chelsea Finn, Peter R. Florence, Chuyuan Fu, Montse Gonzalez Arenas, Keerthana Gopalakrishnan, Kehang Han, Karol Hausman, Alexander Herzog, Jasmine Hsu, Brian Ichter, Alex Irpan, Nikhil J. Joshi, Ryan C. Julian, Dmitry Kalashnikov, Yuheng Kuang, Isabel Leal, Sergey Levine, Henryk Michalewski, Igor Mordatch, Karl Pertsch, Kanishka Rao, Krista Reymann, Michael S. Ryoo, Grecia Salazar, Pannag R. Sanketi, Pierre Sermanet, Jaspiar Singh, Anika Singh, Radu Soricut, Huong Tran, Vincent Vanhoucke, Quan Ho Vuong, Ayzaan Wahid, Stefan Welker, Paul Wohlhart, Ted Xiao, Tianhe Yu and Brianna Zitkovich. "RT-2: Vision-Language-Action Models Transfer Web Knowledge to Robotic Control." ArXiv abs/2307.15818 (2023).

[28] Reed, Scott, Konrad Zolna, Emilio Parisotto, Sergio Gomez Colmenarejo, Alexander Novikov, Gabriel Barth-Maron, Mai Gimenez, Yury Sulsky, Jackie Kay, Jost Tobias Springenberg, Tom Eccles, Jake Bruce, Ali Razavi, Ashley D. Edwards, Nicolas Manfred Otto Heess, Yutian Chen, Raia Hadsell, Oriol Vinyals, Mahyar Bordbar and Nando de Freitas. "A Generalist Agent." Trans. Mach. Learn. Res. 2022 (2022).

[29] Driess, Danny, F. Xia, Mehdi S. M. Sajjadi, Corey Lynch, Aakanksha Chowdhery, Brian Ichter, Ayzaan Wahid, Jonathan Tompson, Quan Ho Vuong, Tianhe Yu, Wenlong Huang, Yevgen Chebotar, Pierre Sermanet, Daniel Duckworth, Sergey Levine, Vincent Vanhoucke,

Karol Hausman, Marc Toussaint, Klaus Greff, Andy Zeng, Igor Mordatch and Peter R. Florence. "PaLM-E: An Embodied Multi-modal Language Model." International Conference on Machine Learning (2023).

[30] Inoue, Y., & Ohashi, H. (2022). Prompter: Utilizing Large Language Model Prompting for a Data Efficient Embodied Instruction Following. ArXiv, abs/2211.03267.

[31] Liu, B., Jiang, Y., Zhang, X., Liu, Q., Zhang, S., Biswas, J., & Stone, P. (2023). LLM+P: Empowering Large Language Models with Optimal Planning Proficiency. ArXiv, abs/2304.11477.

[32] Singh, I., Blukis, V., Mousavian, A., Goyal, A., Xu, D., Tremblay, J., Fox, D., Thomason, J., & Garg, A. (2022). ProgPrompt: Generating Situated Robot Task Plans using Large Language Models. 2023 IEEE International Conference on Robotics and Automation (ICRA), 11523-11530.

[33] Yu, W., Gileadi, N., Fu, C., Kirmani, S., Lee, K., Arenas, M. G., Chiang, H.L., Erez, T., Hasenclever, L., Humplik, J., Ichter, B., Xiao, T., Xu, P., Zeng, A., Zhang, T., Heess, N.M., Sadigh, D., Tan, J., Tassa, Y., & Xia, F. (2023). Language to Rewards for Robotic Skill Synthesis. ArXiv, abs/2306.08647.

[34] Kim, B., Kim, J., Kim, Y., Min, C., & Choi, J. (2023). Context-Aware Planning and Environment-Aware Memory for Instruction Following Embodied Agents. ArXiv, abs/2308.07241.

[35] Huang, W., Xia, F., Xiao, T., Chan, H., Liang, J., Florence, P.R., Zeng, A., Tompson, J., Mordatch, I., Chebotar, Y., Sermanet, P., Brown, N., Jackson, T., Luu, L., Levine, S., Hausman, K., & Ichter, B. (2022). Inner Monologue: Embodied Reasoning through Planning with Language Models. CoArialArialnference on Robot Learning.

[36] Ouyang, L., Wu, J., Jiang, X., Almeida, D., Wainwright, C. L., Mishkin, P., Zhang, C., Agarwal, S., Slama, K., Ray, A., Schulman, J., Hilton, J., Kelton, F., Miller, L.E., Simens, M., Askell, A., Welinder, P., Christiano, P. F., Leike, J., & Lowe, R.J. (2022). Training language models to follow instructions with human feedback. ArXiv, abs/2203.02155.

[37] Xie, Y., Yu, C., Zhu, T., Bai, J., Gong, Z., & Soh, H. (2023). Translating Natural Language to Planning Goals with Large-Language Models. ArXiv, abs/2302.05128.

[38] Shridhar, M., Thomason, J., Gordon, D., Bisk, Y., Han, W., Mottaghi, R., Zettlemoyer, L., & Fox, D. (2019). ALFRED: A Benchmark for Interpreting Grounded Instructions for Everyday Tasks. 2020 IEEE/CVF

Conference on Computer Vision and Pattern Recognition (CVPR), 10737-10746.

[39] Min, S., Chaplot, D.S., Ravikumar, P., Bisk, Y., & Salakhutdinov, R. (2021). FILM: Following Instructions in Language with Modular Methods. ArXiv, abs/2110.07342.

[40] Siciliano B., Khatib O. Handbook of Robotics. 2. izd. Berlin, Heidelberg: Springer-Verlag, 2016. 2304lpp. ISBN 978-3-319-32550-7.

[41] Jeon J., Jung H., Yumbla F., Luong T. A., Moon H. Primitive Action Based Combined Task and Motion Planning for the Service Robot. In: Frontiers in Robotics and AI, 2022, 9.

[42] Gizzi E., Castro M. G., Sinapov J. Creative Problem Solving by Robots Using Action Primitive Discovery. In: 2019 Joint IEEE 9th International Conference on Development and Learning and Epigenetic Robotics (ICDL-EpiRob), 2019, 228-233.

[43] Strudel R., Pashevich A., Kalevatykh I., Laptev I., Sivic J., Schmid C. Learning to combine primitive skills: A step towards versatile robotic manipulation. In: 2020 IEEE International Conference on Robotics and Automation (ICRA), 2019, 4637-4643.

[44] Moeslund T. B., Hilton A., Krüger V. A survey of advances in vision-based human motion capture and analysis. In: Computer Vision and Image Understanding, 2006, 104(2-3), 90-126.

[45] Simeonov A., Du Y., Kim B., Hogan F. R., Tenenbaum J. B., Agrawal P., Rodriguez A. A Long Horizon Planning Framework for Manipulating Rigid Pointcloud Objects. In: *Conference on Robot Learning*, 2020.

[46] Fikes R., Nilsson N. J. STRIPS: A New Approach to the Application of Theorem Proving to Problem Solving. Artificial Intelligence, 1971, 2, 189-208.

[47] McDermott D., Ghallab M., Howe A. E., Knoblock C. A., Ram A., Veloso M. M., Weld D. S., Wilkins D. E. PDDL-the planning domain definition language, 1998.

[48] Lin, K., Agia, C., Migimatsu, T., Pavone, M., Bohg, J. Text2Motion: From Natural Language Instructions to Feasible Plans. In: ArXiv, 2023.

[49] Racinskis P, Arents J, Greitans M. A Motion Capture and Imitation Learning Based Approach to Robot Control. Applied Sciences. 2022; 12(14), 7186.

[50] Jenkins O. C., Matarić M. J. Deriving action and behavior primitives from human motion data. In: IEEE/RSJ International Conference on Intelligent Robots and Systems, 2002, 3, 2551-2556.

[51] Beetz M., Stulp F., Esden-Tempski P., Fedrizzi A., Klank U., Kresse I., Maldonado A., Ruiz F. Generality and legibility in mobile manipulation: Learning skills for routine tasks. In: Autonomous Robots, 2010, 28(1), 21-44.

[52] Sarathy V., Scheutz M. MacGyver problems: Ai challenges for testing resourcefulness and creativity. In: Advances in Cognitive Systems, 2018, 6, 31–44.

[53] Hogan F. R., Ballester J., Dong S., Rodriguez A. Tactile Dexterity: Manipulation Primitives with Tactile Feedback. In: 2020 IEEE International Conference on Robotics and Automation (ICRA), 2020, 8863-8869.

[54] Simeonov A., Du Y., Kim B., Hogan F. R., Tenenbaum J. B., Agrawal P., Rodriguez A. A Long Horizon Planning Framework for Manipulating Rigid Pointcloud Objects. From: Conference on Robot Learning, 2020.

[55] Stone A., Xiao T., Lu Y., Gopalakrishnan K., Lee K., Vuong Q.H., Wohlhart P., Zitkovich B., Xia F., Finn C., Hausman K. Open-World Object Manipulation using Pre-trained Vision-Language Models. In: ArXiv, abs/2303.00905, 2023.

[56] Ugalde F. R. Compact Models of Objects for Skilled Manipulation, 2015.

[57] Mahler J., Matl M., Satish V., Danielczuk M., DeRose B., McKinley S., Goldberg K. Learning ambidextrous robot grasping policies. In: Science Robotics, 2019, 4.

7

An Overview of the Automated Optical Inspection Edge AI Inference System Solutions

Claudio Cantone[1] and Alberto Faro[2]

[1] High Technology Systems H.T.S. srl, Italy
[2] Deepsensing, DEEPS, Italy

Abstract

The aim of this chapter is to provide an overview of automated optical inspection (AOI) edge artificial intelligence (AI) inference system solutions in the digital industry by considering if, and how, they enable manufacturers to reach a satisfactory trade-off between customer needs and production costs. Numerous solutions can address customer and factory needs, from inspection machines to testing boards equipped with cameras installed near the conveyor belt. In all the considered solutions we can implement effective defect detection algorithms, such as the latest You Only Look Once (YOLO) variants based on deep learning (DL), to obtain high key performance indicators (KPIs), i.e., mean average precision, adequate process capability and high throughput yield. Parallel implementations of edge test systems allow us to further improve production yield, while repeated tests performed in sequence can allow us to approach the precision required for zero defect practice. The comparison of available solutions using KPIs, functional requirements (FRs) and non-functional requirements (NFRs) highlights that the advantage of using inspection machines is that they are equipped with user interface and data analysis which helps workers and managers to ensure high quality production process and effective order management. Their weakness is the high cost of purchase and energy consumption, whereas solutions that use

DOI: 10.1201/9788770041027-7   153

computing boards for defect testing at the edge are featured by lower costs. A demonstrator to evaluate the effectiveness of edge AI solutions based on the test boards available on the market and those developed by the EdgeAI project is outlined.

Keywords: automated optical inspection, key performance indicators, functional, non-functional requirements, deep learning, PCB defect detection, edge computing, online and continual learning, process capability.

7.1 Introduction

The advent of cyber physical systems (CPS), i.e., IoT systems equipped with computational capabilities, is affecting the control systems in every industrial and service sector [28], [11]. CPSs allow computational systems to reside ever closer to the production process, reducing latency and increasing throughput yield (TPY), one of the most important KPIs in production processes.

This trend towards edge computing-based inspection systems is particularly evident in the AOI of industrial products. This is our field of research interest in the EdgeAI project.

In this context, we are faced with two different evolutions of the AOI. On the one hand traditional AOI systems that operate at the operational level are being rethought as intelligent systems to be coupled to the production line. On the other hand, CPSs equipped with camera are increasing their computational capacity to achieve effective AOI systems using local or cloud DL algorithms.

The aim of this chapter is to provide an overview of AOI edge AI inference system solutions by discussing if, and how, they allow manufacturers to reach a satisfactory balance between customer needs (mainly in terms of product quality and on-time delivery of ordered lots) and production costs. Section 2 provides the context for us to classify edge AI inference system solutions for AOI, where specific products and prototypes are highlighted to flesh out the discussion. Section 3 compares the leading AOI solutions identified in Section 2 using both KPIs and functional/non-functional requirements. Section 4 outlines the demonstrator we are setting up in the EdgeAI project to improve optical inspection in the digital industry. This will allow us to highlight how edge AI solutions can outperform or complement conventional AI-based inspection machines for AOI.

7.2 Overview of the Main Edge AI Solutions for AOI

Traditional AOI machines are tipically designed to support Surface Mount Technology (SMT) for mounting and interconnecting electronic components on printed circuit boards.

Figure 7.1a outlines the process by which an empty printed circuit board (PCB) is gradually filled with all the components to obtain a fully functional printed circuit board (PCBA). We note that in the figures of this chapter it is assumed that PCBAs are inserted into the AOI system to detect PCB or PCBA defects, i.e. defects relating respectively to the printed circuits or to the component assembly process. Also it should be noted that AOI machines are used online in two stages of the SMT process: at the exit of the Pick and Place process and after the reflow oven to detect almost any surface defect. AOI machines are primarily dedicated to discovering 2D and 3D PCBA defects and making Coordinate Mounting Measurements (CMM). There are multifunctional machines on the market that perform not only AOI and CMM but also Solder Paste Inspection (SPI) [22]. After AOI and X-ray Inspection (AXI) discover surface and internal defects respectively, an electronic test phase consisting of In Circuit Test (ICT) and Functional Test (FCT) is performed.

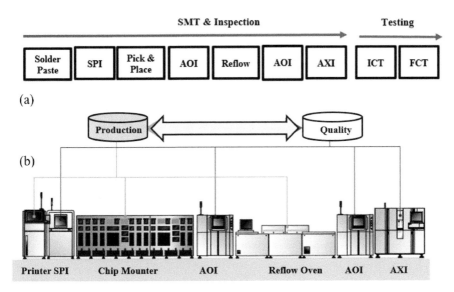

Figure 7.1 a) Printed Circuit Board (PCB) Assembly Process, and b) Typical Implementation of the SMT Production Line, where production data are taken from Printer, Chip Mounter and Reflow, whereas quality data are taken from SPI, AOI and AXI.

An example illustrating this way of using AOI is shown in Figure 7.1b which shows how OMRON proposes to use AOI to discover surface defects using 2D/3D optics and internal defects using X-ray machines [23]. This last control is increasingly widespread, as underlined in [39]. Sometimes the AOI machine is only placed after the reflow oven. In principle this solution is less expensive, although finding defects after reflow oven costs the manufacturer much more to rectify.

The above pattern is followed by mass production factories. In fact, Electronic Manufacturing Services (EMS) factories that produce small manufacturing lots featuring high technology for New Product Introduction (NPI), often adopt offline solutions to avoid changing the path of the conveyor belt in the production site. For this reason, the role of optical inspection machines can be schematized within the production line in two main ways: as control systems not necessarily close to the conveyor belt (Figure 7.2a), and as integrated control systems in the production line (Figure 7.2b) to ensure high production yield, especially in the case of mass production.

Inspection machines have recently been equipped with DL algorithms to improve the accuracy of the defect detection process, such as the Omron VT-S1080. Modern optical inspection machines can be roughly viewed as intelligent edge computing solutions for AOI, whose main problem remains the high purchase and power consumption cost and the constraints they impose on the conveyor belt layout.

A further weakness of inspection machines concerns the AI algorithms used. In fact, if they improve throughput by switching from statistical algorithms to DL inspection-based algorithms, their performance may not reach

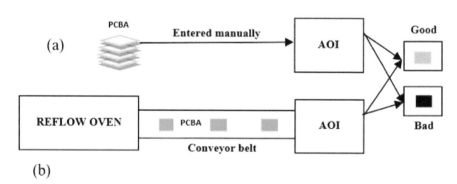

Figure 7.2 Main Inspection Machine Configurations for AOI in the Digital Industry

the high accuracy of 99.8% reported in [23] or the very low rate of defective products reported in [26], if they are not equipped with:

a) Online Learning (OL) to use experimental data to optimize the initial learning model typically obtained from data available in the literature or from similar cases, and

b) Continual learning (CL) to use experimental data to extend the learning capacity of the algorithm to discover further defects without forgetting the previous ones.

Therefore in this chapter, by OL and CL we mean a learning technique that uses experimental data to improve defect detection accuracy and to learn additional defects respectively. Regarding OL, we have to note that in the global industry, online learning should be used with caution if a test program executed on one site is expected to produce the same results as that implemented on the other sites [18]. This implies that global manufacturers should check whether OL improves defect inspection equally across their different sites. In this case, or in the case of tests on local production lines, it is useful for the learning model to be continuously updated from the images captured by the cameras to optimize the discovery of defects or to deal with different types of defects.

In principle, inspection machines can be equipped with OL and CL, but this will increase their cost as it requires the machines to be equipped with a powerful processing CPU or powered by an additional GPU due to the high computational load required by such algorithms [27].

For this reason, AOI solutions have recently appeared on the market consisting of powerful workstations, possibly equipped with GPU boards, and equipped with a high-resolution camera installed near the belt, such as those proposed in [7] using the Neousys technology (Figure 7.3a) and those proposed by Advantech [2], ADLINK [8] and AAEON [1]. Advantech and AAEON solutions are shown in Figure 7.3b and 7.3c.

The hardware architectures shown in Figure 7.3 allow us to highlight that the Neousys and AAEON solutions use a powerful workstation able of both training and testing, while Advantech uses a processing unit for testing and a GPU workstation for learning and for the classification of defects. ADLINK's solution can be achieved by replacing MIC-720 with EOS-I6000-M which is an AI vision system suitable for testing and classification, while learning takes place on cloud server.

Although these solutions support both online and continual learning and real-time verification of product defects, the problem remains of the high

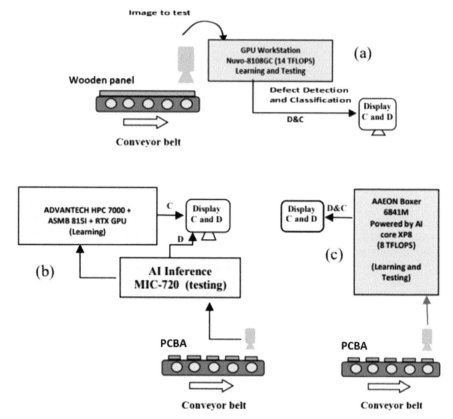

Figure 7.3   Edge AI AOI solutions from Neousys (a), Advantech (b), and AAEON (c) for Defect Detection (D) and Classification (C). The model is Pre-trained on the Workstation.

purchase and power consumption cost, as well as a certain difficulty of installing such systems near to the production line due to their size and conditioning constraints.

Alternatively, a solution where visual testing is done at the edge and learning in the cloud can reduce purchase and power consumption costs without increasing latency, as shown in Figure 7.4. This solution can be obtained by replacing the MIC-720 unit with a NVIDIA board in the Advantech proposal shown in Figure 7.3. In Figure 7.4 a Jetson board is adopted for edge tests, for example Jetson TX2 as proposed in [29]. In the latter case, learning is on the cloud but several tests suitable for highlighting groups of defects can be performed in parallel by competing boards thus decreasing latency.

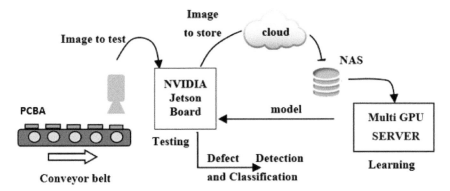

Figure 7.4 AOI Solution Consisting of an Edge Board for Testing and a GPU Server for Learning.

7.3 Comparing EdgeAI solutions for AOI using Relevant KPIs, NFRs and FRs in Digital Industry

In the previous section we outlined three main EdgeAI solutions for AOI, namely the one based on an inspection machine (see Figure 7.1), hereinafter referred to as IS, the one that makes use of a GPU workstation equipped with high-precision cameras (see Figure 7.3a and 7.3c), called GS, and the one based on a test board near the conveyor belt that sends images to the cloud server for online and continual learning (see figures 7.3b and 7.4), called ES. In the chapter we also consider a fourth solution consisting of cameras sending images to a cloud server for testing and learning, which we will call CS.

The discussion of such solutions was mainly based on cost and flexibility aspects and suggested to take into great consideration both CS and ES. In this section, we compare these solutions by considering Key Performance Indicators, Functional and Non Functional Requirements.

7.3.1 Comparison using KPIs

KPIs mainly deal with cost effectiveness, efficiency (precision) of the discrimination process and its productivity (speediness) as suggested in [13] to evaluate the performance of every digital system. Efficiency of the discovery process is usually evaluated, as in every information retrieval system, using precision and recall that may be easily obtained by the confusion matrix related to the adopted discovery algorithm [15]. The confusion matrix together with the precision and recall formulas are shown in Figure 7.5.

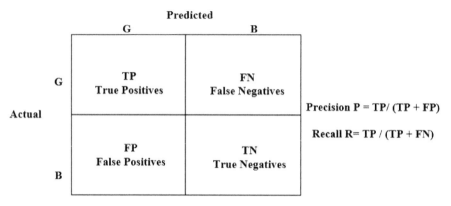

Figure 7.5   Confusion Matrix and Precision/Recall Formulas

In some cases, the most important characteristic is recall, for example, if we are interested in finding out all, or almost all, defective PCBAs, it is reasonable to increase the false positive checking effort, while in other cases it may be better to use precision, for example, if one is interesting that the discrimination process only outlines not defective PCBAs although high accuracy may increase false negatives.

At first glance, one might think that precision may be the most important feature in AOI of PCBAs, especially in cases where the requirement for near-zero defects should be adopted, for example in the aerospace sector or recently in the automotive industry. But precision alone can cause many good products to be discarded, thus increasing production costs. For this reason, efficiency indicators that combine precision and recall are used in the digital industry such as the mAP defined as the mean of the average precisions, and the F-measure defined as the weighted harmonic mean of precision and recall. The following balanced F-measure is often used, denoted as F1, which equally weighs precision and recall:

$$F1 = 2 \cdot P \cdot R / (P+R)$$

We use mAP as it is widely adopted to measure the performance of defect detection algorithms in industrial manufacturing. The meaning of mAP can be understood by introducing the notion of Intersection Of Union (IoU) [37], a measure from 0 to 1 of the similarity between the bounding box containing a possible defect and the one relating to a real one (the ground truth). According to [37] IoU is used as a threshold for whether an object having a defect-like image should enter the defective class (i.e., class consisting of defective

PCBAs) or not. Other rules can be found in the literature for whether a PCBA defect should be predicted as real, for example, in [32] a possible defect contained in a bounding box is predicted as a real defect if both IoU and another coefficient, namely the confidence coefficient, calculated by the detection algorithm are greater than 0.5.

Thus, choosing a high IoU will increase the percentage of really good items compared to those predicted good by the algorithm, but even numerous really good items may be discarded (that is, false negatives increase). Conversely, a low IoU will decrease false negatives, but defect discovery is characterized by low precision thus increasing false positives as the chosen similarity is not sufficient to discriminate good from bad elements. Consequently, to reduce the AOI alarms for possible false positives, e.g., the PCBAs featured by IoU > 0,5 and whose confidence coefficient is close to 0,5, it is advisable to increase precision by adopting a most performing algorithm or to increase IoU since this generally implies an increase of the confidence too, even this is not desired since it implies an increase in false negatives.

The mAP is obtained by evaluating the average precision of the controls performed for each IoU value from 0.5 to 0.9 with a step of 0.1 and performing the mean of these averages [32]. To simplify, in the work we will use $mAP_{0.5}$, i.e. the precision of the discovery process for IoU = 0.5. Therefore $mAP_{0.5}$ = 0.99 does not mean that we will have 1% error, but that the error of the predicted good items is close to 1% with a reasonably low number of false negatives, i.e., few good products will be discarded from the ones for customers.

The above considerations justify why the efficiency of the discovery process is evaluated using mAP. We recall that the mAP depends not only on the efficiency of the discovery algorithm but also on the type of defect to be found. Typical defects to be discovered on the PCB are missing hole, mouse bite, open circuit, short circuit, spurious copper, spur. But measurements of the relevant metrological data of the PCBA are also useful, such as component height, lift, tilt, missing or incorrect component, incorrect polarity, flipped component, OCR inspection of 2D code, component offset (X / Y/rotation), fillet (e.g., end joint width, wetting angle, side joint length), exposed zone, foreign material, zone error, cable offset, cable posture, cable presence, sphere of weld, weld bridge, distance between components and component angle.

Several algorithms have been proposed in the literature to manage the problems listed above. A study highlighting different algorithms to manage either PCB or PCBA defects can be found in [12] where it is demonstrated that $mAP_{0.5}$ ranges from 95% to 98%. In this chapter we update this study considering the best performing algorithms for typical PCB defect detection. From the literature we found that these are mainly optimized versions of the DL-powered YOLO algorithm [35]. The $mAP_{0.5}$ of such algorithms increased from 95.7% proposed in 2018 [5] to higher values using the best performing DL algorithms developed from 2018 to present. For example, $mAP_{0.5}$ is 99% in the algorithm proposed in [38], 99% in [25], 98.7% in [36], 99% in [39]. Such values go beyond 99% more recently, i.e. 99.17% in [20], 99.5% in [11] and 99.71% in [40].

Although this comparison has only an indicative value since the mentioned precision values were not achieved using the same data set [33], we can reasonably assume that the solutions denoted with IS, GS and CS can be equipped with a DL algorithm whose defect discovery precision could increase from 98% to 99.7%, and that this could be further improved by online learning to 99.8%, as stated in [23].

The feasibility of implementing YOLO-based algorithms on ES has been recently shown in literature thus confirming that ES can also be equipped with such an algorithm, e.g., in [30] a YOLO implementation on NVIDIA Jetson TX2 is illustrated in characterized by satisfactory precision performance, that is, $mAP_{0.5} = 98\%$. We are currently working on solving two open problems: a) to what extent more accurate algorithms can be implemented on ES and b) how to implement such algorithms on less expensive boards (e.g., Jetson Nano and Raspberry PI4) by extending the DL algorithms proposed in [14] and [34].

However, although the theoretical accuracy of the optimized YOLO versions has reached a very high value, it may not be sufficient for the quality control of PCBAs to be used in applications where the constraint of near-zero defects is required, such as in the automotive industry [4]. In fact, 99.8% of $mAP_{0.5}$ approximately implies that the delivered defective products are about 2000 per million, whereas 1000 per million defective parts is a typical expected value in automotive products satisfying the near-zero defect constraint [24].

Note that the former failure rate, known as defects per million opportunities (DPMO) [16], measures all PCBA possible failures, i.e., defects of components or due to the assembly process. If each PCBA consists of approximately 100 components, this means that the DPMO is 2000 defective PCBAs per million if the PCB is filled with components with a failure rate of 20ppm. The DPMO in the industrial sector is used as a relevant KPI to measure the process capability, i.e., how well the process yield meets customer expectations in terms of acceptable defective products. This capability can also be expressed by a percentage (called Yield) or by a coefficient named Cpk, i.e., a statistical coefficient between 0 and 2 where Cpk = 2 means that there are no defective PCBAs leaving the production process, while Cpk = 0 means that the quality process does not detect any fault, so all the faulty boards are still in the leaving products. A conversion table is available in the literature to pass from DPMO to Yield or to Cpk and vice versa, e.g., in [31].

In the semiconductor industry, DPMO = 6000 is an acceptable value if the near-zero defect constraint is not required. Using the conversion table, we can find that this corresponds to Cpk = 1.33 and Yield = 99.40%. Consequently, if we aim to have DPMO = 6000 for both defective components and surface defects, using the conversion table we obtain that we must use 99.55% of non-defective components plus an instrument, such as AOI for example, obtaining 99. 80% accuracy to discriminate between good and bad products coming out of the SMT process. The latter accuracy in detecting surface defects can be achieved by recent versions of the AI-YOLO algorithm, but applications characterized by the zero-defect constraint require a Yield of 99.98% which can be achieved using an AOI of 99.95% of precision.

Therefore, while waiting for more performing algorithms, it is reasonable to carry out two or three repetitions of the AOI checks of the products classified as good to improve the accuracy as proposed in [9]. Indeed, this is a reasonable procedure only if the AOI checks are statistically independent. as claimed in [6] due to the noise superimposed on the images when they are taken by the cameras, for example due to faded colours or weaknesses in the lighting system. Consequently, reproducing the control using the same AOI can eliminate the uncertainty due to noise issues thus allowing the AOI to approach its maximum theoretical accuracy calculated using literature data.

Also, as pointed out earlier, one could reduce the volume of bad products delivered as good (i.e., to reduce false positives) by increasing the IoU, but this usually also increases false negatives. In fact, this can lead to consider good products the ones that are close to the boundary between the "good" and "bad" classes and which are affected by the maximum uncertainty of classification.

Therefore, checking the PCB using another AOI system, i.e., not repro-ducing the measurement but replicating it using a different AOI, can be useful to improve accuracy without changing the recall. This repetition could add some defective elements to the "bad class" as suggested in [21] which, hopefully, could coincide with the few defective products that were not detected by the first test. This is also stated in [9] where it is underlined

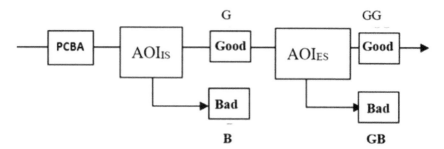

Figure 7.6   Repeating the AOI Check.

that repetition improves accuracy in the electronics industry even if beyond a certain threshold repetition is not cost effective due to the increasing cost of adding a further check.

This consideration suggests evaluating in our project the possibility of adding a check after the inspection machine using an ES check to try to satisfy the constraint of near-zero defects. In fact, the hypothesis of adding a further control to the one currently carried out without modifying the layout is an opportunity given the low cost and the high flexibility of the ES. Figure 7.6 shows how the repetition scheme proposed in [9] can be reworked to improve the mAP of AOI. Test repetition to avoid bad products reaching customers could be carried out, even manually, only for testing the few PCBAs that passed the first test but were classified close to the border between good and bad clusters.

In addition to mAP and process capability to evaluate process efficiency, another important KPI is the productivity of an AOI system, also known as throughput yield (TPY), to measure good PCBAs at optical control output in the unit of time. Since in all considered IS, GS and CS the test phase is performed on GPU machines, the comparison can be made considering the latency due to the algorithm and the camera system used to acquire the images of the PCBAs on the belt. Latency mainly depends on the implementation of the algorithm and is often not indicated in the literature. It can be measured indirectly by the speed, in frames per second (FPS), at which the proposed algorithm is able to process the images.

A general comparison of the FPS achievable in the available methods for defect discovery can be found in [30] where the authors pointed out that their version of the YOLO algorithm is able to reach about 90 FPS. This value is also confirmed in other studies, for example we found that the FPS goes from 33 FPS in [40] to 90 FPS in [20]. Lower but satisfactory FPS characterize

3D defect detection, for example 19 FPS in [Du, 2023]. Regarding ES, we found from the literature that even in ES the implementation of YOLO algorithms can achieve high throughput, for example, in [30] it is proved that the DL-based YOLO algorithm implemented on Jetson TX2 can process 22 FPS [30].

Therefore, using a TX2 board, it can be expected that a 25 cm^2 PCB can be inspected in about 90 msec, if each image taken by the camera is about 5 cm^2. This means that the AOI production per hour obtainable using ES could be around 3250 boards per hour (bph) which is a value comparable with the value of 4189 bph obtained using IS reported in [26]. Let us note that such values refer to the number of PCBA exiting from the optical inspection phase (see fig.1.1a) . Indeed other electrical checks may decrease such thoughput, e.g., the ones dealing with the determination of the safe operating area of PCBs to be used in power applications. In [26] the authors proposed other relevant KPIs beyond hourly production, i.e., precision of detected defects, working time and delivery times from order to shipment.

The accuracy of defect discovery can be calculated using mAP as shown above, while the last two proposed KPIs depend on the organization of work. Therefore, they can only be analysed by knowing the factory organization structure, order volume and rate. It is out of the scope of the chapter. However, they suggest us that mAP and FPS alone are not sufficient to measure the impact of AOI on the SMT process. In fact, cycle time and takt time should also be included in the KPIs at least to verify that the AOI production system can meet the time constraints due to customer orders. A general discussion may be found in [17]. For the paper, it is sufficient to include the following parameters in the KPI list:

- Cycle time (CT), i.e., the time required to produce a lot of PCBs requested by the customer divided by the number of PCBs.
- Takt time (TT), i.e., the time interval during which the production line is available in the time interval required by the customer to produce the PCB lot divided by the number of PCBs to be delivered to the customer. In other words, it is the maximum time interval for producing one PCBA to meet the customer time constraint considering the availability of the production resources and the number of PCBAs of the lot.

Knowing CT and TT we can verify the condition necessary to satisfy the customer's demand, i.e., CT < TT. This means that CT (the inverse of throughput yield) is a very important KPI that should be appropriately scaled down to meet overall customer demand in due time. This can be achieved: i) by increasing the FPS of the AOI unit, ii) by using more than one AOI unit in parallel, or iii) by implementing more than one production line. The first two conditions can be obtained more conveniently by ES than by IS since its low cost allows adopting many cameras to work in parallel. In fact, the CT of a production line can be improved by passing from a solution in which a camera sends images to a testing board as illustrated in Figure 7.7a. to the one proposed in [3] made up of several cameras possibly equipped with a testing board (Figure 7.7b). In both cases, the images are sent to a server to update the pre-trained model.

To get an idea of the cost savings using ES in both cases illustrated in Figure 7.7 let us consider the market cost of CS, ES, and IS. Assuming one CS as a unit cost, from the market cost we found that this cost becomes 2 for ES, 6 for a WS provided with GPUs and from 25 to 50 for IS depending on if the IS is a low-cost machine or a professional one. Therefore, the purchase costs are as follows: n+6 for CS, 2n+6 for ES and 25 or 50 for IS where n is the number of cameras and related testing boards.

Using these values, Figure 7.8a compares the costs of ES and CS with the cost of a low cost IS proposed by Saki in [29]. This comparison is feasible since they have the same configuration, i.e., they are all equipped with a camera which, thanks to a telecentric lens system (Figure 7.7a), takes pictures of PCB slices of about 5 x 25 cm while it is placed on the conveyor belt.

Instead, to evaluate the cost savings by using multiple cameras and boards, we compare the CS and ES with the OMRON professional solution, i.e., VT-S1080, assuming that the CS (ES) is equipped with 5 camera positions (5 camera positions plus 5 testing boards) as in Figure 7.7b so that the whole PCB can be inspected as it is transported on the conveyor belt and at the same time the OMRON AOI captures all images of the PCB using a robotic system that moves the camera over the PCB inside the machine. The comparison is shown in Figure 7.8b.

Figure 7.8 clearly shows that in both cases CS and ES are less expensive than IS. The cycle time using CS and ES in case 3.3a. is greater than that of IS, then CS and ES are suggested only if a relatively high CT is acceptable. If a lower CT is required, the parallel implementation is recommended. Furthermore, we should mention that both mAP and FPS could be further improved in ES by using more performant testing boards like the ones proposed in [19] where it is stated that object classification can be performed at hundreds of FPS. This is for further study.

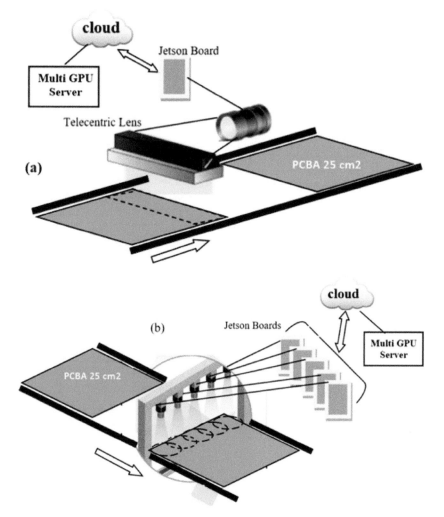

Figure 7.7 a) A Camera Equipped with a Testing Board Which Sends the Image of a PCB Slice of about 5 X 25 cm Using a Telecentric Lens to a Testing Board, b) A Set of Five Cameras Equipped with Testing Boards. Images Are Sent to a Server to Update the Pre-Trained Model. The Server Periodically Sends the Updated Model to the Edge Testing Boards.

7.3.2 Comparison using NFRs

In addition to the mentioned KPIs, further quality requirements, so-called non-functional requirements, should be considered to compare the different solutions. Below we indicate some NFRs that consider those

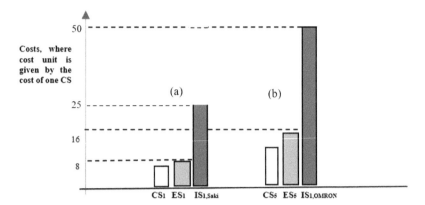

Figure 7.8 An Approximate Comparison of the Purchase Costs of CSs and ESs Equipped with one Camera Versus the Low Cost 2D Saki AOI (a), and the Purchase Costs of CSs and ES Equipped with Five Cameras Seats Versus the Professional 2D/3D OMRON AOI (b).

proposed in [26] for the support of workers, i.e., the adopted solution should:

- Enable efficient use of workers' time through automation.
- Improve control capability through real-time data feedback.
- Explain the defects at least by locating the defects found on the PCB or PCBA, which allows workers to improve the production process.

All the above NFRs can be satisfied by CS and ES provided that appropriate user interfaces are developed that help workers interpret and manage data from optical inspection.

NFRs are proposed in [26] dealing with planning tools, i.e., managers should be supported by suitable planning tools, based on data from AOI and other IoT monitoring systems, to meet takt time and to verify more generally that the overall time including the purchase of the raw material and the delivery of the products to customers (i.e., the lead time) is compatible with the customer's demand. This implies that lead time should also be included in the KPI list above.

Since the current IS and GS provide effective operator interfaces and planning tools for managers, these solutions, despite the high cost, can maintain some advantage over ES until ES is equipped with the mentioned worker interface and management tools.

This can be facilitated by the fact that ESs can take full advantage of parallel technology and cloud computing. In fact, the worker interface, usually installed near the production line, could be implemented by adding another board to the edge testing system, while data analysis tools could be implemented on the network server available to plant managers.

This issue should be addressed carefully in ES and is a challenge for any project aiming to use AOI testing boards at the edge.

7.3.3 Comparison using functional requirements

To complete the comparison, the main FRs must also be considered. The following list consists of five FRs, of which the first two are mandatory while the last three are highly recommended. Such FRs require that any AI solution for AOI:

a) it should have a high mAP suitable to support the process capability required by the industrial sector of interest of the producers, for example CPk = 1.33 for the semiconductor industry. Based on the discussion in this section, all solutions could meet this requirement using OL-based DL defect detection algorithms.

b) it should be able to detect PCB defects in real time as the PCBs are transported on a conveyor belt. Based on the discussion made in this section, all solutions can meet this requirement due to their relatively high FPS value.

c) it must have an adequate feedback loop with the machine controls. This requirement also belongs to the NFR list mentioned, but here it is understood as the requirement that the solution has a minimum set of functions to help workers and managers optimize the PCB production. IS and GS usually satisfy such FR, while it is acceptable for CS and ES to provide at least some defect location functionality to explain the causes of the defect.

d) it should learn to discover defects by exploiting the data available in the literature. In principle, this requirement is satisfied by GS, CS and ES as they are usually open systems, while ISs are usually designed as closed solutions which are not provided with network attached storage system on the cloud to include data from the literature to improve the accuracy of learning or to handle new defects.

7.3.4 Advantages of ES with respect to the other approaches

The comparison of the AI based AOI solutions using the main KPIs, NFR and FR has highlighted that ES is a promising technology provided it is equipped with adequate operational and management interfaces.

Some points that encourage the effort to equip ESs with such interfaces are not only their low cost and parallel processing that allow them to achieve better KPIs for the detection of multiple defects simultaneously, but also the possibility for ESs to take full advantage of the cloud technology not only to use the cloud to better build the mentioned user interfaces, but also to enable small companies to use AOI-based control remotely.

7.4 Edge AI Solutions Demonstrator

Given the different solutions available for optical defect detection, a demonstrator can be useful to evaluate if and how an AOI solution can help in practice manufacturers to reach a satisfactory compromise between the quality required by customers (in terms of acceptable number of defective items and takt time) and factory costs.

Currently we are activating such demonstrator equipped with the following technologies:

- An IS machine, supplied by HTS, i.e., OMRON AI-AOI VT-S1080, to measure mAP and FPS achievable during the PCBA test and to verify that it is able of achieving using Deep Learning the high accuracy required by the industrial sector of interest, i.e., semiconductor or automotive sectors.

- A workstation, provided by DEEPS, equipped with a 7 TB storage system, an AMD Ryzen^{TM} Threadripper 3970 CPU and two NVIDIA RTX 6000 GPUs. This workstation currently acts as a server on the local network so it will allow us to simulate CS ed ES but could be connected via a fast channel to cameras to simulate GS as well.

- Several NVIDIA boards, namely Jetson Nano, Jetson TX2, Xavier and ORIN, to host the algorithm trained on the GPU server at the edge and a NAS (Network Attached Storage) system to store the images taken by the cameras to allow the server to online update the pre-trained model.

- High-resolution Basler cameras to take images of PCBs as they are being transported on a conveyor belt. These images will be sent to the server's NAS or workstation near the belt or passed through a fast channel to the NVIDIA cards.

The components from b) to d) will allow us to activate the demonstrator using the same platform illustrated in Figure 7.4, to measure the most significant KPIs and to evaluate the NFR for defect detection. In this way, commercially available ES and new EdgeAI AOI solutions, such as the one based on the low-power board to be developed by the EdgeAI project, could be compared with GS, CS and IS.

We note that the main purpose of the demonstrator is not to support designers in developing DL defect detection algorithms that outperform the current ones, even if this test can also be performed using the platform, but to demonstrate that: a) the DL-based defect models obtained from the pre-training phase on the server can be implemented on the edge boards to obtain test performance comparable to that of IS and GS but at a lower cost as required mainly by mass production companies, and b) the AI Edge solution can be equipped with extremely precise defect discovery and defect explainability algorithms to support the improvement of the production process and in the identification of possible critical components as required mainly by NPI.

Furthermore, the conditions suggesting the combination of different solutions can be studied. For example, if a low throughput yield is acceptable, this may justify CS over the others. In addition, an ES-based remote solution will be tested to support small companies in implementing a simple and cost-effective solution where the testing board is installed close to the conveyor belt and the learning powered by OL and CL is done by a cloud server.

7.5 Conclusion

An overview of the available solutions for AI-based optical defect inspection of PCBAs has been made from an engineering point of view, i.e., emphasizing whether and how they can support a satisfactory trade-off between product quality and production costs.

From the overview it emerged that it is possible to adopt different solutions to meet the needs of the factory and customers, from inspection machines to testing boards equipped with cameras installed near the conveyor belt. Generally, in all the considered solutions it is possible to implement effective defect detection algorithms, such as the latest DL-based YOLO versions, to obtain the suitable mean precision, i.e., mAP, to support the required process capability.

The main advantage of using inspection machines is that they have data analysis tools that support managers to ensure high quality and effective management planning. Their weakness, i.e., the high cost of purchase and energy consumption, is the strength of solutions that use processing boards for defect testing at the edge.

Parallel implementations of edge solutions, using suitable optical systems, improve latency and the number of PCBAs that may be classified as good or bad products per time unit, while repeated tests carried out by a test board installed after the inspection machine, allow us to approach the process capability required in industry sectors characterized by the near zero defects constrain. This can be achieved without decreasing recall, thus avoiding an increase in false negatives.

It was discussed how a solution can achieve a low cycle time that can meet takt time and lead time to satisfy customer demand, emphasizing that using ES this can be achieved by increasing the FPS of the AOI and activating, if necessary, parallel AOI units in the production line.

A suitable platform was also presented to evaluate the most suitable solutions using experimental data. This will help us demonstrate the efficiency, productivity, and cost-effectiveness of a solution in practice and test whether coprocessing units, such as the recent neuromorphic boards, can improve discovery algorithms. It will allow us also to demonstrate how small companies can use the platform to perform defect detection using local testing boards supervised by a remote server.

Acknowledgements

This research was conducted as part of the EdgeAI "Edge AI Technologies for Optimised Performance Embedded Processing" project, which has received funding from KDT JU under grant agreement No 101097300. The KDT JU receives support from the European Union's Horizon Europe research and innovation program and Austria, Belgium, France, Greece, Italy, Latvia, Luxembourg, Netherlands, and Norway.

References

[1] AAEON, "AI@Edge: AI Vision in Automated Optical Inspection", 2023, Available at:https://newdata.aaeon.com.tw/DOWNLOAD/application/BOXER-6841M%20AOI%20Story.pdf

[2] Advantech, "AI AOI Selected Success Story", 2023, Available at:https://advcloudfiles.advantech.com/membership/upload/ee3b76aa/AI-AOI Success-Story-CollectionFinal.pdf

[3] Advantech, "Building AOI Technology to Accelerate Yield in precision manufacturing", March 2023, Available at:https://www.advantech.com/en-eu/resources/case-study/building-aoi-technology-to-accelerate-yield-in-precision-manufacturing

[4] AEC-Q004: Zero Defects Guideline. Automotive Electronics Council, 2020, Available at:http://www.aecouncil.com/Documents/AEC_Q00 4_Rev-.pdf

[5] L, Chang, "A CNN Based Reference Comparison Method for Classifying Bare PCB Defects", *The Journal of Engineering*, 2018 (16)

[6] S. Chan, "How Repeatable Is Your AOI? An Easy and Quick Way To Find Out", 2017, Available at:http://www.linkedin.com/pulse/how-rep eatable-your-aoi-easy-quick-way-find-out-steven-chan

[7] L.C. Chen, M.S. Pardeshi, WT Lo, et al., "Edge-glued wooden panel defect detection using deep learning". *Wood Science Technology* **56**, 477–507 (2022).

[8] Y. Chia-Wei, "How Edge AI Can Improve the Visual Inspection Process", *Quality Magazine*, September 30, 2020

[9] Y. Chung-Huang, C. Jwu-E, "Application of Three-Repetition Tests Scheme to Improve Integrated Circuits Test Quality to Near-Zero Defect", *Sensors* 22 (4158), May 2022

[10] A. Costanzo, A. Faro, D. Giordano and C. Spampinato, "An ontological ubiquitous city information platform provided with Cyber-Physical-Social-Systems", *2016 13th IEEE Annual Consumer Communications & Networking Conference (CCNC)*, Las Vegas, NV, USA, 2016, pp. 137-144. Available at:https://doi.org/10.1109/CCNC.2016.7444746

[11] Y. Du, et al., "An automated optical inspection (AOI) platform for three-dimensional (3D) defects detection on glass micro-optical components (GMOC)", *Optics Communications*, to appear on Volume 545, 15 October 2023

[12] A.A.R.M.A. Ebayyeh and A. Mousavi, "A Review and Analysis of Automatic Optical Inspection and Quality Monitoring Methods in Electronics Industry", *IEEE Access*, vol. 8, pp. 183192-183271, 2020

[13] Eurocontrol, KPI Drafting Group, "Cost Effectiveness and Productivity KPIs", October 2001, Available at:http://www.eurocontrol.int/sites/defa ult/files/2019-05/cost-effectiveness-and-productivity-kpis-2001.pdf

[14] R. Faro, "Object Detection and Semantic Segmentation models for Defect Detection in Wood Production", Undergraduate thesis, Department of Electrical, Electronics and Computer Engineering University of Catania, 2023

[15] T, Fawcett, "An Introduction to ROC Analysis", *Pattern Recognition Letters*, 27 (8): 861–874. 2006

[16] K. Feldeman, "Driving Quality Improvement with DPMO: A Roadmap to Process Excellence", July 17, 2023 Available at:https://www.sixsigma.com/dictionary/defects-per-million-opportunities-dpmo/

[17] E. Fogg, "Takt Time vs. Cycle Time vs. Lead Time: definitions and calculations", *Machine Metrics, Manufacturing Analytics*, September 24, 2020

[18] D, Haigh, "Repeatability Is AOI Watchword for Volume SMT Production", 1999, Available at:http://www.edn.com/repeatability-is-aoi-watchword-for-volume-smt-production/

[19] Hailo, "Delivering Unprecedented Performance to a Diverse Range of Edge AI Applications", 3 August 2023, Available at:https://www.edge-ai-vision.com/companies/hailo/

[20] J.Y. Lim, J.Y.1 Lim, V.M. Baskaran, X. Wang, "A deep context learning based PCB defect detection model with anomalous trend alarming system", *Results in Engineering*, Volume 17, March 2023

[21] R.J Mackenzie, "Repeatability vs. Reproducibility", *Technology Networks*, March 25, 2019

[22] Nordson, SQ3000^{TM}+ multi-Function for 3D AOI, SPI & CMM, online 2021

[23] OMRON, "VT-S1080, Next generation 3D AOI", 2023, Available at:https://inspection.omron.eu/en/products/vt-s1080#features

[24] R. Oshiro (2018), Fundamentals of AEC-Q100: "What Automotive Qualified Really Means", Available at:https://media.monolithicpower.com/mps_cms_document/w/e/Webinar_Fundamentals_-_of_AEC-Q100-6Nov2018.pdf

[25] J.H. Park, Y.S. Kim, H. Seo H, Y.J. Cho. "Analysis of Training Deep Learning Models for PCB Defect Detection", *Sensors* (Basel). 2023 Mar 2;23(5):2766

[26] M. Park, J. Jeong, "Design and Implementation of Machine Vision-Based Quality Inspection System in Mask Manufacturing Process", *Sustainability* 2022, 14, 6009. Available at:https://doi.org/10.3390/su14106009

[27] A. Prabhu, et al, "Computationally Budgeted Continual Learning: What Does Matter?", *Proceedings of the IEEE/CVF Conference on Computer Vision and Pattern Recognition* (CVPR), 2023, pp. 3698-3707

[28] M. Ryalat, H. El Moaqet, M. Al Faouri, "Design of a Smart Factory Based on Cyber-Physical Systems and Internet of Things towards Industry 4.0". *Appl. Sci.* 2023, *13*, 2156. Available at:https://doi.org/10.3390/app13042156

[29] Saki, "2D AOI technology", Available at:https://www.sakicorp.com/en/company/technology/2daoi_tec/

[30] S. Shaojun, J. Junfeng, H. Yanqing, S. Mingyang, EfficientDet for fabric defect detection based on edge computing, *Journal of Engineered Fibers and Fabrics* Volume 16: 1–13, 2021.

[31] Six Sigma Material, "Conversion Tables", Available at:https://www.six-sigma-material.com/Tables.html

[32] J. Solawetz, "What is Mean Average Precision (mAP) in Object Detection?" Roboflow, May 6, 2020, Available at:https://blog.roboflow.com/mean-average-precision/

[33] G, Spadaro, et al., "Towards One-Shot PCB Defect Detection with YOLO", Ital-IA 2023: *3rd National Conf. on Artificial Intelligence*, organized 000by CINI, Pisa, IT May 29–31, 2023

[34] A. Strano, "Chip Surface Defect Classification: A Benchmark Analysis of Deep Learning Architectures", Undergraduate thesis, Department of Electrical, Electronics and Computer Engineering, University of Catania, 2023

[35] J. Terven, D. Cordova-Esparza, "A Comprehensive Review of YOLO: From YOLOv1 and Beyond", August, 2023, Under review in ACM Computing Survey, Available at:https://doi.org/10.48550/arXiv.2304.00501

[36] Y. Yang, K. Haiyan, "An Enhanced Detection Method of PCB Defect Based on Improved YOLOv7", *Electronics* 12, no. 9: 2023 Available at:https://doi.org/10.3390/electronics12092120

[37] S. Yohanandan, "What is Mean Average Precision (MAP) and how does it work", Xailent, June 5, 2020, Available at:https://xailient.com/blog/what-is-mean-average-precision-and-how-does-it-work/

[38] C. Zhang, W. Shi, X. Li, H. Zhang, H. Liu, "Improved bare PCB defect detection approach based on deep feature learning", *The 2nd 2018 Asian Conference on Artificial Intelligence Technology*, (ACAIT 2018), *J. Eng.*, 2018, Vol. 2018 Is. 16, pp. 1415-1420, 2018.

[39] Q. Zhang, et al., "Deep learning-based solder joint defect detection on industrial printed circuit board X-ray images", *Complex & Intelligent Systems* 8:1525–1537, 2022

[40] H. Zhu, L. Xing, H. Fan, T. Wu, "New PCB Defect Identification and Classification Method Combining Mobile Net Algorithm and Improved YOLOv4 Model", Research Square, 2022

8

Efficient AI-based Attack Detection Methods for Sensitive Edge Devices and Systems

Daniel Hirsch[1], Falk Hoffmann[1], Andrija Neskovic[2], Celine Thermann[2], Rainer Buchty[2], Mladen Berekovic[2], and Saleh Mulhem[2]

[1]NXP Semiconductors, Germany.
[2]Universität zu Lübeck, Germany

Abstract

An increasing number of edge devices store and process sensitive user data, presenting an attractive target for attackers. This trend of data storage and processing at the edge is expected to continue. As secure devices are integrated into new systems with increased device operation times, exposure to environmental stress also increases significantly. Especially, for standalone micro-Edge devices the relevance of this is increasing. Enhanced protection mechanisms are required and AI-based approaches are promising candidates.

In this contribution, we examine the requirements for such mechanisms and the sensing capabilities of state-of-the-art secure devices. Based on these capabilities and attack models, a dataset for training and validation is generated. Considering the requirements and the available dataset, a selection of applicable algorithms is defined. The selected algorithms are evaluated and compared based on the obtained results and computational loads, as the basis for future work.

Keywords: artificial intelligence, machine learning, security, attack detection, edge AI, micro-Edge, autonomous security, AI security.

DOI: 10.1201/9788770041027-8   177

8.1 Introduction and Background

Edge Computing (EC) is one of the most practical computing concepts used in day-to-day life applications. The architecture of edge computing is illustrated in Figure 8.1. EC is divided into three levels: Edge/IoT device, Edge device/node, and Cloud level. The core idea of EC is to perform computations and storage directly at the end-user level [1]. i.e., at the network edge [2, 3]. Handling sensitive data becomes prominent. The extraction of these sensitive data and the manipulation of security-relevant features of IoT and edge devices represent a lucrative target for attackers. Therefore, the need to securely protect and handle these data becomes important. Protection mechanisms that work towards this goal can be deployed on all three levels.

Security features are usually implemented to protect these data assets. The most straightforward way to identify manipulation or attacks is by checking both environmental and the device's internal sensors. Another way is to observe the logical monitoring and protection mechanisms that trigger a device reset or limit further use of the device. In extreme cases, device operation is temporarily or permanently blocked.

False alarms may be triggered in cases where sensor information is directly used without any further evaluation of severity, application relevance,

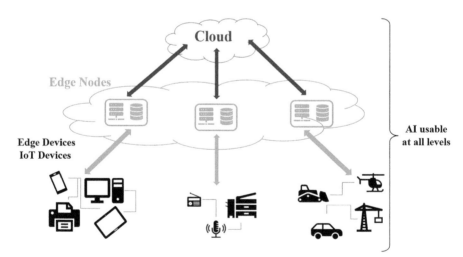

Figure 8.1   Architecture of Edge System

or statistical analysis of environmental effects. The consequence of such false alarms may be severe, leading to DoS attacks, for example. As secure devices are being integrated into an increasing number of systems with extended duty cycles, even up to permanent power-on conditions, the exposure to environmental stress increases significantly. A more advanced evaluation of sensor events and more flexible reactions need to be considered.

To ensure correct functionality of these systems and the integrity of user data, the evaluation of the security mechanisms with the help of AI algorithms represents a promising alternative to conventional approaches. To identify applicable algorithms for attack detection, an evaluation of the requirement specifications is carried out. However, due to the field of application and the special limitations in the physical domain of IoT and other edge devices, the requirements are challenging.

Relevant attacks on the edge

Studying security attacks on electronic devices and systems is a well-established field, but edge devices have certain characteristics that make them more prone to certain attacks and threats when compared to more capable computing devices. In [4], some of the aspects are pointed out, namely:

- **Weak Computation Power:** Edge devices are less powerful than cloud servers, making them susceptible to attacks not effective on cloud counterparts. Fragile defence systems on edge devices further expose them to unique threats.
- **Attack Unawareness**: IoT's lack of user interfaces limits awareness of device status, hindering attack detection.
- **Operating System (OS) and Protocol Differences:** Edge devices lack uniform OSes and protocols, complicating the creation of a unified security approach.
- **Limited Access Control Precision:** Edge computing's complex systems demand fine-grained access control, unlike current coarse-grained models.

Figure 8.2 shows distributions of the attack types on edge devices as presented in [4].

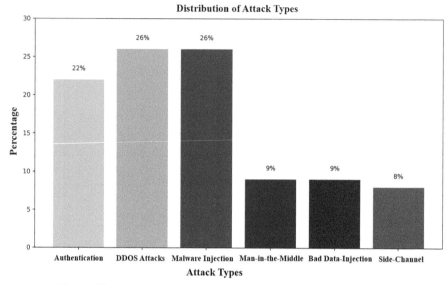

Figure 8.2   Possible Attacks against Edge Devices (Adapted From [4])

In the following, we summarize some possible attacks against edge devices, cloud, and edge systems:

1. Possible Attacks on Edge Devices and Nodes

- **Malware Injection Attacks:** Malware Injection Attacks inject malicious code into the target device. These attacks can lead to arbitrary code execution which can compromise the security of further devices in the network. Considering the case of edge devices, protection becomes much more difficult because the limited computing power does not allow for classical high-performance firewalls or threat protection systems, like with general-purpose computers.

- **DDOS Attacks:** DDoS, short for Distributed Denial of Service, is a cyber assault that involves perpetrators attempting to interrupt the regular operations of one or multiple servers. This is achieved by leveraging distributed resources, often in the form of a network of compromised edge devices, also known as a botnet [4]. It constitutes a potent form of attack that seeks to hinder the legitimate utilization of a particular service.

- **Authentication and Authorization Attacks:** Authentication is the processing of verifying a user's identity who requests certain services and authorization grants that user rights to perform operations.

An adversary could exploit weaknesses in the authentication and authorization mechanisms to obtain privileged access rights and perform malicious operations.

- **Side-channel Attacks:** Refer to a type of attack, where the adversary can exploit information leakage of security-sensitive information via publicly accessible information which is not security-sensitive by nature. Most prominent examples of side-channel attacks exploit the power consumption or timing behaviour of a device while executing sensitive information. Side-channel attacks on the device level can potentially come from two sources, malicious tasks or a malicious OS. Task-level attacks or Timing attacks are typically cache-based such as Flush+Reload [5], Flush+Flush [6], Prime+Probe [7], Evict & Time [8], Evict & Reload [9], Spectre [10] and Meltdown [11] attacks. Here, the attacker aims at getting sensitive data by exploiting sharing vulnerabilities in caches [10, 11] and, in the case of Spectre and Meltdown, out-of-order optimization issues. The attack surface is large, also comprising several proposed and existing covert-channel attacks [12, 13, 14]. Multiple mitigation techniques have already been proposed, typically featuring either logical or physical separation, noise-based techniques, scheduler-based techniques, and constant time techniques. Attackers can exploit the power consumption of the edge device via a power side-channel attack. The concept of side-channel analysis appeared in the late 1990s [15], with Differential Power Analysis (DPA) [16] becoming a successful attack method. It was utilized to attack AES with a Simple Power Analysis (SPA) [17]. With the growing interest in the topic, more elaborate attack methods have been presented, e.g., Correlation Power Analysis (CPA) [18]. These types of attacks pose a significant threat to security-critical applications. Nowadays, even more powerful attack methods based on Template Attacks or utilizing AI as an attack tool for side-channel analysis are present. Fault injection attacks aim at maliciously altering an edge device's functionality. This can range from disturbances in the power supply voltage, irregularities in the clock signal, electromagnetic or radiation disturbances or overheating as described in [19]. The attack objective could be as complex as revealing the secret key of cryptographic primitives, but also simple, like blocking the computation, i.e., denial of service. The complexity and cost to perform a successful Fault Injection Attack can vary based on equipment costs and required knowledge about the underlying hardware. Although fault attack can be expensive in terms of complexity and cost, it is practical and can be mounted on most commonly used architectures from ARM, Intel and AMD [20]. In recent years, even more elegant software-based approaches exploiting voltage scaling led to successful attacks on the Intel SGX secure enclave [21].

2. Possible Attacks on the Cloud

Securing cloud services is mainly achieved by separating a cloud's tenants. These must not be able to escape their individual virtual machines and get access to other tenant's data. Unfortunately, such has been proven viable via side-channel attacks, leading to cross-VM secret leakage via different levels of CPU cache side-channel attacks [5, 9, 22, 23, 24, 25, 26]. Mitigation is technically possible, but typically requires significant changes to hardware [27, 28, 29, 30], hypervisors [31, 32, 33, 34, 35, 36], or guest OSes [36]. Such approaches are not easily applicable to existing data centres. Mitigation by frequent VM migration [37, 38] is theoretically also feasible but comes at prohibitively high migration cost, i.e. several minutes of migration time [39], and hence only addresses the issue of long-term co-location. Attacks by malicious VMs however take only milliseconds [5, 24].

3. Intrusion Attacks on Edge System

The network-based exchange of data and commands between edge devices and cloud infrastructure implies several threats that can affect the edge system due to an insecure network. For instance, attackers can block the data transfer by malicious gateway access or network floods [40]. Similarly, attackers can perform attacks such as impersonation attacks, communication interception, password guessing attacks, data integrity violations, Denial of Service (DOS) and bad Quality of Service (QoS) [40].

Several countermeasures have been proposed to prevent or detect such attacks. The problem with these existing countermeasures is that they usually only address one specific attack, where an attacker can launch a multitude of attacks. To identify such attacks, two essential approaches exist which are signature-based and anomaly-based detection. By nature, signature-based attacks can be overcome by altering the attack code to evade detection and do not protect against previously unknown attacks [41]. Anomaly-based detection, in turn, is prone to false positives as legitimate applications may appear as malicious [41]. The combination of both methods mitigates some of the named individual shortcomings [41, 42].

Intrusion detection systems (IDS) are essential tools for monitoring and detecting of anomalous activities in a network of edge devices and systems and responding to these attacks. Traditional IDS relies on signatures or rules to detect known attacks, but these methods are not effective against new and evolving threats. An anomaly detection system, on the other hand, relies on identifying abnormal behaviour in network traffic data. However, these systems can generate false positives, making them less reliable, and are inability to detect new/unknown attacks [40].

8.2 Efficient Attack Detection

In this section, the approach of selecting an appropriate solution for attack detection on resource-constrained micro-Edge ICs is described. The standalone IC protects on-chip data and secrets by preventing unauthorized access. First, the requirements related to this task are described, followed by a section on the dataset. Based on the requirements and the available dataset, a selection of applicable algorithms obtained from thorough research is specified.

8.2.1 Requirements

An implementation must meet requirements in the three domains of security, user experience, and realizability. The correlation of the requirements focusing on the three general domains is depicted in Figure 8.3 using a top-down representation.

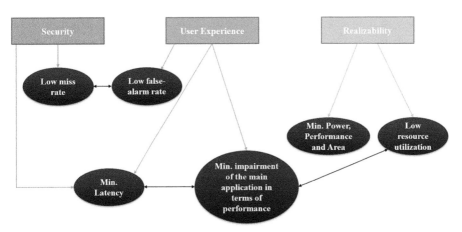

Figure 8.3   Correlation of requirements

The main goal is to target the domain of security. Based on AI, an algorithm capable of improving the present security mechanisms will be researched and evaluated. Since the devices handle sensitive data during operation, requirements covering the targeted levels of security must be defined. The second domain is represented by the user experience. Due to the commercial nature of the products, and since the additional functionality does not necessarily translate to a direct added value for the user, the user experience during usage is not allowed to be negatively influenced by the implemented solution.

Lastly, it must be noted that the available resources for implementing the functionality on the considered devices are limited in terms of computational power, area, and current consumption. Therefore, to benefit from the developed solution, it is also necessary to formulate implementation requirements that are realistic and applicable. These requirements are summarized in the domain of realizability.

Starting from the security perspective, the target of evaluation can contain highly sensitive data, therefore, a low miss rate in terms of detection of actual and exploitable attacks is mandatory to ensure the security and integrity of data stored on the device. Also, the implemented solution should not reduce the usability of the product or affect the user experience negatively. This requirement demands the lowest possible false alarm rate. Furthermore, the implemented solution should have minimal impact on the performance of the main application to achieve a satisfying user experience.

Besides the presented requirements, a fast response time constitutes a very important requirement in this application. To react quickly and prevent performance issues, the response time needs to be as fast as possible. This requirement can be attributed to the domains of security and user experience. From a security perspective, a fast response time is required to protect the secrets stored on the device. From the user's perspective, customers are not keen to see longer response times when using the devices. Therefore, in both domains, a fast response time is seen as advantageous.

Since the developed solution is targeted to be implemented on low-power edge or IoT devices, the available resources are very limited. Based on these general preconditions, further requirements concerning the memory, required die area, power efficiency, and CPU usage need to be formulated. Especially considering the CPU usage, low utilization must be achieved to guarantee minimal impairment of the main application.

8.2.2 Underlying Dataset

To obtain a flexible solution that is applicable for a variety of devices, the detection capabilities of state-of-the-art devices will be investigated. Since a dataset cannot be obtained from measuring traces in the laboratory or gathering field data, the dataset must be generated artificially. For this purpose, the relevant phases within an application and available inputs will be analysed.

Based on this further possible attack scenarios need to be researched and modeled. By the combination of capabilities and the theoretical consideration of attacks, a dataset will be derived. In the process of dataset generation, reasoned assumptions must be made and all decisions must be evaluated critically. Furthermore, the choice of labeling is going to be justified and strategies for the generation of a subset for the model validation will be explained.

8.2.3 State-of- the-Art Attack Detection Methods

1. AI-based Attack Detection at Edge Device Level
   The use of AI methods provides efficient countermeasures. HAL [43] provides a quantitative and qualitative analysis of several machine-learning models for use in cache-based side-channel attack detection. It specifically addresses real-time requirements, detection at an early stage, and minimal performance overhead and demonstrates this in the context of security applications (RSA and AES cryptosystems). It however does not provide a definite answer on specific model usage.
   Similarly, WHISPER [44] proposes a tool for side-channel attack detection based on machine learning. Instead of using a single approach, it features multiple ML models in combination that interpret the behavioural data of concurrent processes. This data is collected via hardware performance counters. The authors demonstrate the tool's capability by achieving >99% accuracy of detecting a large and diverse attack vector while introducing only a reasonably low performance overhead.
   In today's secured systems, installation and execution of malicious application software is typically rendered impossible by so-called shielded execution e.g. provided by the Intel Software Execution Guard [45]. However, such shielded execution can be compromised by privileged attackers, e.g. by changing page-table entries of memory pages that are specifically used by shielded execution. By this approach, a malicious OS kernel can observe corresponding memory-page accesses

and hence extract potentially sensitive information. DejaVu [46] is a software framework that enables self-protection detecting such privileged side-channel attacks from within the shielded execution. This is enabled by the so-called pathlet execution time. For this, a dedicated reference clock is employed that is specifically constructed using the Intel Transactional Synchronisation Extension (TSX). By featuring this robust reference clock, not only deviations in pathlet execution time indicating an attack can be detected but also interruptions of the reference-clock thread resulting in a transaction timeout.

Modern processors provide a limited number of registers known as hardware performance counters (HPCs) that capture hardware-related events. These special-purpose registers can be used to study the impact of side-channel attacks (SCAs). Compared to normal operation, the number of events when a system is under attack appears noticeably different. [47] explores several different machine-learning models for real-time cache-based SCA detection using HPCs. 16 HPC features are collected for both victims under attack and victims not under attack at different sampling rates. Overhead is reduced, by only using four features. This way they all can be fetched synchronously. The authors determined that for a sampling granularity of $500\,\mu s$, the systems incur 5% overhead while maintaining good detection accuracy. In addition, they also considered the latency for the different models. It was found that the Decision Tree provides the best trade-off between performance and latency.

[48] provides another approach using HPCs for the detection of side-channel attacks. The authors provide a two-step process comprised of an offline and an online phase. In addition to covering cache-based SCAs, they also consider branch-based and DRAM-based SCAs. The HPC can be observed to follow a Gaussian distribution with different means and variances. Anomalous behaviour shows a different distribution with a different mean. During the offline phase, data is collected in different environments. This includes benign programs running in the background, that make intensive use of the cache, branching, or the RAM. Afterwards, during the online phase, HPCs are collected and classified using an AI model. To counteract the high number of false-positives, anomalous traces are correlated with traces in a database. A high correlation indicates an attack, while a low correlation indicates a benign program running.

2. AI-based Attack Detection at Cloud Level

   Several security countermeasures have been introduced to mitigate possible attacks on the cloud [49, 50]. For instance, CloudRadar [41] proposes an approach to secure the cloud. This approach correlates signature-based and anomaly-based detection techniques in to spot side-channel attacks. Here, signature-based detection is used to identify when a protected VM executes cryptographic applications. Anomaly-based detection is orthogonally used to monitor and identify abnormal cache behaviours typical of cache-based side-channel attacks. As such, the approach is non-intrusive, not requiring any changes to hardware, hypervisor, guest VM, and applications. It hence is comparatively easy to deploy in existing cloud environments and, according to the authors, requires only patching and a little overhead [51]. To improve the performance of such detection techniques and cover more than the classical cache attacks against edge devices, Recurrent Neural Networks (RNNs) were proposed in [52]. The results show that additionally to the classical detection of cache attacks, the RNN-based solutions efficiently detect Rowhammer, Spectre, Meltdown, and Zombieload attacks as well.

3. AI-based Intrusion Detection System for Edge Systems

   AI-based Intrusion Detection Systems (AI-IDS) are a promising alternative to traditional IDS. AI methods can identify patterns and anomalies in network traffic data, enabling it to detect previously unseen, unknown, and complex threats. AI-IDS faces three main challenges: (1) the quality of the data used for training and testing the models, (2) the accuracy of the chosen AI algorithm, and (3), the performance of the chosen AI algorithm.

   Various techniques have been proposed to enhance the accuracy and performance of AI-IDS. For instance, the use of sampling techniques to select representative datasets can improve both the accuracy and speed of intrusion detection [53]. By combining a sampling technique with a random forest machine learning algorithm, IDS exhibits very good performance. However, it shows also different levels of detection accuracy for different attacks. In [54], Gini Impurity-based Weighted Random Forest (GIWRF) was used as a data feature selection technique. Then, the accuracy of several AI algorithms deployed as AI-IDS was analyzed. The results show that AI accuracy ranges from 88.99% to 99.98%.

8.2.4 Selection of Applicable Algorithms

In the following, the algorithms from research are evaluated in terms of their applicability to the problem with the associated requirements.

- Neural Network (NN): Model built from basic computation units called neurons that are usually organized into layers. Connections between neurons are associated with trainable weights. Upon receiving input, the input is weighted and aggregated. Afterwards, a possibly non-linear function is applied. The complexity of these models increases with the number of layers [55]. A Perceptron [56] is the simplest possible model and consists of a single layer of neurons. In contrast, Multilayer feed-forward Networks are comprised of multiple layers that are connected in a feed-forward fashion. If there are not only forward connections but also those connecting neurons to previous layers, the network is called recurrent [55]. An example of these types of networks are long short-term memory (LSTM) networks. These networks contain memory cells, making it possible to retain information [57].
- Trees: Models that make their decisions based on tree-like structures. One example of these types of models are decision trees (DT). They can be used for both classification and regression tasks [55]. Isolation Forests on the other side aim to find anomalies using binary trees [58].
- Support Vector Machine (SVM): Algorithm that tries to find a separator with the maximum distance to training samples. In the simplest case, the goal is to find a simple linear separator between two classes in a two-dimensional space [55].
- Bayesian Network: Probabilistic model allowing for computation of posterior probability distributions. Nodes represent random variables, while edges describe conditional dependencies between variables. Each node is associated with some probabilities that quantify the effect on other nodes. These probabilities can be learned from a given dataset. The simplest example of these classifiers are Naive Bayes classifiers [55].
- Instance-based: Algorithms that directly estimate from a given dataset. Processing of the input is deferred until queried. After answering the request, all intermediate results are discarded [55]. The most well-known example of these algorithms is the k-nearest neighbour (KNN) algorithm [59].
- Linear Regression: Algorithms that try to find the best-fitting function for some given data. In the simplest case, the goal is to find a linear function for a single input variable. Depending on the application, more complex functions might be used [55].

- Discriminant Analysis: Methods aiming to estimate the decision boundary between classes. Approaches like linear discriminant analysis might make simplifying assumptions, such as an underlying Gaussian distribution for all classes and the same covariance matrices for all classes [60].
- Ensemble: Combining multiple algorithms to achieve a better outcome. A random forest (RF) is an ensemble of decision trees. Ensembles can be created by many different techniques. One such technique is called boosting. It aims at improving performance by assigning higher weights to examples that have been misclassified and thus making an incentive to classify them correctly for the next model in the ensemble [55].

The most limiting factor in the selection of a suitable algorithm comes in the form of resource limitations. Some models have significant requirements for the systems they are executed on. Examples of such models are Neural Networks that can easily have millions of parameters. Not only does this require sufficient storage, but might also cause a significant delay in loading and applying these parameters. Consequently, a separate accelerator might be required, that increases the area consumed. Even non-parametric algorithms like KNN might not be a good solution, as the whole dataset has to be stored. Depending on the size of the dataset, this might also put a significant strain on the amount of memory available.

In [44], the results for twelve different machine learning models were presented, covering all of the classes described above. Considering all models achieving 80% accuracy leaves SVMs, DTs, RFs, KNN, NNs and Ensemble learning. As discussed beforehand, both KNN and NNs have high computational requirements, making them not suitable for the application.

Due to the experimental setup in [44], the detection latency using the models is unknown on the Edge and is to be determined in the future. Thus, the impact of deploying them cannot be determined, and there is still a need for AI models that offer (1) high detection accuracy, (2) efficiency, and (3) meet the requirements of Edge devices. Such an AI model can serve as a highly accurate, efficient and lightweight attack detector at the edge level.

8.3 Discussion and Conclusion

The continuously growing use and uptime of secure devices increase the number of sensor events during the product lifetime. Especially, for standalone micro-edge devices this becomes more relevant. Consequently, this is pushing the industry to step up from direct reaction to sensor events towards more advanced solutions. It is, however, paramount that such solutions do not negatively influence both device operation and user experience. Processing and interpretation of available sensor information with the help of artificial intelligence offers the possibility to develop future solutions.

AI-based approaches particularly overcome limitations of established solutions based on signatures and anomaly detection: Signature-based approaches are inherently limited to known attacks and their signatures. They hence neither provide protection against future attacks nor altered attack code. Approaches based on anomaly detection, in turn, are prone to false positives as legitimate, non-malicious code may trigger such detection. So far, AI methods have been successfully employed in a wide variety of security systems, covering both edge nodes and cloud environments. They provide a sufficiently high detection rate at minimal false-positive level and, by nature, are immune to evasion strategies like altered attack code. However, so far no single gold solution exists. For AI approaches the choice of a suitable AI method is paramount. Similarly, sufficient labelling strategies and derived training sets need to be developed. Finding an optimal AI strategy for a given threat scenario is hence still open to research.

Acknowledgement

This research was conducted as part of the project *"Edge AI Technologies for Optimised Performance Embedded Processing"* (EdgeAI), which has received funding from KDT JU under grant agreement No 101097300. The KDT JU receives support from the European Union's Horizon Europe research and innovation program and Austria, Belgium, France, Greece, Italy, Latvia, Luxembourg, Netherlands, and Norway.

References

[1] H. Xue, B. Huang, M. Qin, H. Zhou and H. Yang, "Edge Computing for Internet of Things: A Survey", 2020 International Conferences on Internet of Things (iThings) and IEEE Green Computing and Communications (GreenCom) and IEEE Cyber, Physical and Social Computing (CPSCom) and IEEE Smart Data (SmartData) and IEEE Congress on Cybermatics (Cybermatics), pp. 755–760, 2020. W442W7302

[2] P. G. Lopez, A. Montresor, D. Epema, A. Datta, T. Higashino, A. Iamnitchi, M. Barcellos, P. Felber and E. Riviere, "Edge-centric Computing: Vision and Challenges", SIGCOMM Comput. Commun. Rev., vol. 45, no. 5, p. 37–42, 2015. W442W7302

[3] W. Shi and S. Dustdar, "The Promise of Edge Computing", IEEE Computer, vol. 49, no. 5, pp. 78–81, 2016. W442W7302

[4] Y. Xiao, Y. Jia, C. Liu, X. Cheng, J. Yu and W. Lv, "Edge Computing Security: State of the Art and Challenges", Proceedings of the IEEE, vol. 107, no. 8, pp. 1608–1631, 2019. W442W7302

[5] Y. Yarom and K. Falkner, "FLUSH+RELOAD: A High Resolution, Low Noise, L3 Cache Side-Channel Attack", Proceedings of the 23rd USENIX Conference on Security Symposium, p. 719–732, 2014. W442W7302

[6] D. Gruss, C. Maurice, K. Wagner and S. Mangard, "Flush+Flush: A Fast and Stealthy Cache Attack", Proceedings of the 13th International Conference on Detection of Intrusions and Malware, and Vulnerability Assessment, vol. 9721, p. 279–299, 2016. W442W7302

[7] M. S. Inci, B. Gulmezoglu, G. Irazoqui, T. Eisenbarth and B. Sunar, "Cache Attacks Enable Bulk Key Recovery on the Cloud", Cryptographic Hardware and Embedded Systems – CHES 2016, pp. 368-388, 2016. W442W7302

[8] D. A. Osvik, A. Shamir and E. Tromer, "Cache Attacks and Countermeasures: The Case of AES", Topics in Cryptology – CT-RSA 2006, pp. 1–20, 2006. W442W7302

[9] D. Gruss, R. Spreitzer and S. Mangard, "Cache Template Attacks: Automating Attacks on Inclusive Last-Level Caches", 24th USENIX Security Symposium (USENIX Security 15), pp. 897–912, 2015. W442W7302

[10] P. Kocher, J. Horn, A. Fogh, D. Genkin, D. Gruss, W. Haas, M. Hamburg, M. Lipp, S. Mangard, T. Prescher, M. Schwarz and Y. Yarom, "Spectre Attacks: Exploiting Speculative Execution", 2019 IEEE Symposium on Security and Privacy (SP), pp. 1–19, 2019. W442W7302

[11] M. Lipp, M. Schwarz, D. Gruss, T. Prescher, W. Haas, A. Fogh, J. Horn, S. Mangard, P. Kocher, D. Genkin, Y. Yarom and M. Hamburg, "Meltdown: Reading Kernel Memory from User Space", 27th USENIX Security Symposium (USENIX Security 18), pp. 973–990, 2018. W442W7302

[12] Y. Lyu and P. Mishra, "A Survey of Side-Channel Attacks on Caches and Countermeasures", Journal of Hardware and Systems Security, vol. 2, no. 1, pp. 33–50, 2018. W442W7302

[13] J. Van Bulck, M. Minkin, O. Weisse, D. Genkin, B. Kasikci, F. Piessens, M. Silberstein, T. F. Wenisch, Y. Yarom and R. Strackx, "Foreshadow: Extracting the Keys to the Intel SGX Kingdom with

Transient Out-of-Order Execution", 27th USENIX Security Symposium (USENIX Security 18), 2018. W442W7302

[14] D. Genkin, L. Valenta and Y. Yarom, "May the Fourth Be With You: A Microarchitectural Side Channel Attack on Several Real-World Applications of Curve25519", Proceedings of the 2017 ACM SIGSAC Conference on Computer and Communications Security, p. 845–858, 2017. W442W7302

[15] R. Mayer-Sommer, "Smartly analyzing the simplicity and the power of simple power analysis on smartcards", International Workshop on Cryptographic Hardware and Embedded Systems, pp. 78–92, 2000. W442W7302

[16] P. Kocher, J. Jaffe and B. Jun, "Differential power analysis", Advances in Cryptology—CRYPTO'99: 19th Annual International Cryptology Conference Santa Barbara, California, USA, August 15–19, 1999 Proceedings 19, pp. 388–397, 1999. W442W7302

[17] S. Mangard, "A simple power-analysis (SPA) attack on implementations of the AES key expansion", Information Security and Cryptology—ICISC 2002: 5th International Conference Seoul, Korea, November 28–29, 2002 Revised Papers 5, pp. 343–358, 2003. W442W7302

[18] E. Brier, C. Clavier and F. Olivier, "Correlation power analysis with a leakage model", Cryptographic Hardware and Embedded Systems-CHES 2004: 6th International Workshop Cambridge, MA, USA, August 11-13, 2004. Proceedings 6, pp. 16–29, 2004. W442W7302

[19] A. Barenghi, L. Breveglieri, I. Koren and D. Naccache, "Fault Injection Attacks on Cryptographic Devices: Theory, Practice, and Countermeasures", Proceedings of the IEEE, vol. 100, no. 11, pp. 3056–3076, 2012. W442W7302

[20] J. Breier and X. Hou, "How Practical Are Fault Injection Attacks, Really?", IEEE Access, vol. 10, pp. 113122–113130, 2022. W442W7302

[21] K. Murdock, D. Oswald, F. D. Garcia, J. Van Bulck, D. Gruss and F. Piessens, "Plundervolt: Software-based Fault Injection Attacks against Intel SGX", 2020 IEEE Symposium on Security and Privacy (SP), pp. 1466–1482, 2020. W442W7302

[22] G. Irazoqui, T. Eisenbarth and B. Sunar, "S $ A: A shared cache attack that works across cores and defies VM sandboxing–and its application to AES", 2015 IEEE Symposium on Security and Privacy, pp. 591–604, 2015. W442W7302

[23] G. Irazoqui, M. S. Inci, T. Eisenbarth and B. Sunar, "Wait a minute! A fast, Cross-VM attack on AES", Research in Attacks, Intrusions and Defenses: 17th International Symposium, RAID 2014, Gothenburg, Sweden, September 17-19, 2014. Proceedings 17, pp. 299–319, 2014. W442W7302

[24] F. Liu, Y. Yarom, Q. Ge, G. Heiser and R. B. Lee, "Last-level cache side-channel attacks are practical", 2015 IEEE symposium on security and privacy, pp. 605–622, 2015. W442W7302

[25] Y. Zhang, A. Juels, M. K. Reiter and T. Ristenpart, "Cross-VM side channels and their use to extract private keys", Proceedings of the 2012 ACM conference on Computer and communications security, pp. 305–316, 2012. W442W7302

[26] Y. Zhang, A. Juels, M. K. Reiter and T. Ristenpart, "Cross-tenant side-channel attacks in PaaS clouds", Proceedings of the 2014 ACM SIGSAC Conference on Computer and Communications Security, pp. 990–1003, 2014. W442W7302

[27] L. Domnitser, A. Jaleel, J. Loew, N. Abu-Ghazaleh and D. Ponomarev, "Non-monopolizable caches: Low-complexity mitigation of cache side channel attacks", ACM Transactions on Architecture and Code Optimization (TACO), vol. 8, no. 4, pp. 1–21, 2012. W442W7302

[28] F. Liu and R. B. Lee, "Random fill cache architecture", 2014 47th Annual IEEE/ACM International Symposium on Microarchitecture, pp. 203–215, 2014. W442W7302

[29] Z. Wang and R. B. Lee, "A novel cache architecture with enhanced performance and security", 2008 41st IEEE/ACM International Symposium on Microarchitecture, pp. 83-93, 2008. W442W7302

[30] Z. Wang and R. B. Lee, "New cache designs for thwarting software cache-based side channel attacks", Proceedings of the 34th annual international symposium on Computer architecture, pp. 494–505, 2007. W442W7302

[31] T. Kim, M. Peinado and G. Mainar-Ruiz, "STEALTHMEM System-Level Protection Against Cache-Based Side Channel Attacks in the Cloud", 21st USENIX Security Symposium (USENIX Security 12), pp. 189–204, 2012. W442W7302

[32] P. Li, D. Gao and M. K. Reiter, "Stopwatch: a cloud architecture for timing channel mitigation", ACM Transactions on Information and System Security (TISSEC), vol. 17, no. 2, pp. 1–28, 2014. W442W7302

[33] J. Shi, X. Song, H. Chen and B. Zang, "Limiting cache-based side-channel in multi-tenant cloud using dynamic page coloring", 2011

IEEE/IFIP 41st International Conference on Dependable Systems and Networks Workshops (DSN-W), pp. 194–199, 2011. W442W7302

[34] V. Varadarajan, T. Ristenpart und M. Swift, "Scheduler-based defenses against Cross-VM side-channels", 23rd USENIX security symposium (USENIX security 14), pp. 687–702, 2014. W442W7302

[35] B. C. Vattikonda, S. Das and H. Shacham, "Eliminating fine grained timers in Xen", Proceedings of the 3rd ACM workshop on Cloud computing security workshop, pp. 41–46, 2011. W442W7302

[36] Y. Zhang und M. K. Reiter, "Düppel: Retrofitting commodity operating systems to mitigate cache side channels in the cloud", Proceedings of the 2013 ACM SIGSAC conference on Computer & communications security, pp. 827–838, 2013. W442W7302

[37] S.-J. Moon, V. Sekar and M. K. Reiter, "Nomad: Mitigating arbitrary cloud side channels via provider-assisted migration", Proceedings of the 22nd acm sigsac conference on computer and communications security, pp. 1595–1606, 2015. W442W7302

[38] Y. Zhang, M. Li, K. Bai, M. Yu and W. Zang, "Incentive compatible moving target defense against vm-colocation attacks in clouds", Information Security and Privacy Research: 27th IFIP TC 11 Information Security and Privacy Conference, SEC 2012, Heraklion, Crete, Greece, June 4-6, 2012. Proceedings 27, pp. 388–399, 2012. W442W7302

[39] V. Varadarajan, Y. Zhang, T. Ristenpart and M. Swift, "A Placement Vulnerability Study in Multi-Tenant Public Clouds", 24th USENIX Security Symposium (USENIX Security 15), pp. 913–928, 2015. W442W7302

[40] N. Chaabouni, M. Mosbah, A. Zemmari, C. Sauvignac and P. Faruki, "Network intrusion detection for IoT security based on learning techniques", IEEE Communications Surveys & Tutorials, vol. 21, no. 3, pp. 2671–2701, 2019. W442W7302

[41] T. Zhang, Y. Zhang and R. B. Lee, "CloudRadar: A Real-Time Side-Channel Attack Detection System in Clouds", Research in Attacks, Intrusions, and Defenses. RAID 2016., pp. 118–140, 2016. W442W7302

[42] M. Alam, S. Bhattacharya, D. Mukhopadhyay and S. Bhattacharya, "Performance Counters to Rescue: A Machine Learning based safeguard against Micro-architectural Side-Channel-Attacks", IACR Cryptol. ePrint Arch., 2017. W442W7302

[43] M. Mushtaq, A. Akram, M. K. Bhatti, M. Chaudhry, M. Yousaf, U. Farooq, V. Lapotre and G. Gogniat, "Machine learning for security: The case of side-channel attack detection at run-time", 2018 25th IEEE

International Conference on Electronics, Circuits and Systems (ICECS), pp. 485–488, 2018. W442W7302

[44] M. Mushtaq, J. Bricq, M. K. Bhatti, A. Akram, V. Lapotre, G. Gogniat and P. Benoit, "WHISPER: A tool for run-time detection of side-channel attacks", IEEE Access, vol. 8, pp. 83871–83900, 2020. W442W7302

[45] O. Aciiçmez, "Yet another microarchitectural attack: exploiting I-cache", Proceedings of the 2007 ACM workshop on Computer security architecture, pp. 11–18, 2007. W442W7302

[46] S. Chen, X. Zhang, M. K. Reiter and Y. Zhang, "Detecting privileged side-channel attacks in shielded execution with Déjá Vu", Proceedings of the 2017 ACM on Asia Conference on Computer and Communications Security, pp. 7–18, 2017. W442W7302

[47] H. Wang, H. Sayadi, A. Sasan, S. Rafatirad, T. Mohsenin und H. Homayoun, "Comprehensive Evaluation of Machine Learning Countermeasures for Detecting Microarchitectural Side-Channel Attacks", Proceedings of the 2020 on Great Lakes Symposium on VLSI, pp. 181–186, 2020. W442W7302

[48] M. Alam, S. Bhattacharya and D. Mukhopadhyay, "Victims Can Be Saviors: A Machine Learning–Based Detection for Micro-Architectural Side-Channel Attacks", J. Emerg. Technol. Comput. Syst., vol. 17, no. 2, 2021. W442W7302

[49] S. Briongos, G. Irazoqui, P. Malagón and T. Eisenbarth, "Cacheshield: Detecting cache attacks through self-observation", Proceedings of the Eighth ACM Conference on Data and Application Security and Privacy, pp. 224–235, 2018. W442W7302

[50] M. Chiappetta, E. Savas and C. Yilmaz, "Real time detection of cache-based side-channel attacks using hardware performance counters", Applied Soft Computing, vol. 49, pp. 1162–1174, 2016. W442W7302

[51] Z. Liu, B. Xu, B. Cheng, X. Hu and M. Darbandi, "Intrusion detection systems in the cloud computing: A comprehensive and deep literature review", Concurrency and Computation: Practice and Experience, vol. 34, 2021. W442W7302

[52] B. Gulmezoglu, A. Moghimi, T. Eisenbarth and B. Sunar, "Fortuneteller: Predicting microarchitectural attacks via unsupervised deep learning", arXiv preprint arXiv:1907.03651, 2019. W442W7302

[53] J. Ren, J. Guo, W. Qian, H. Yuan, X. Hao and H. Jingjing, "Building an effective intrusion detection system by using hybrid data optimization based on machine learning algorithms", Security and communication networks, 2019. W442W7302

[54] R. A. Disha und S. Waheed, "Performance analysis of machine learning models for intrusion detection system using Gini Impurity-based Weighted Random Forest (GIWRF) feature selection technique", Cybersecurity, Bd. 5, Nr. 1, p. 1, 2022. W442W7302

[55] S. Russel and P. Norvig, Artificial Intelligence: A Modern Approach, Prentice Hall, 2010. W442W7302

[56] D. Rosenblatt, "The perceptron: A perceiving and recognizing automaton", Cornell Aeronautical Laboratory, 1957. W442W7302

[57] S. Hochreiter and J. Schmidhuber, "Long short-term memory", Neural computation, vol. 9, no. 8, pp. 1735–1780, 1997. W442W7302

[58] F. T. Liu, K. M. Ting and Z.-H. Zhou, "Isolation forest", 2008 eighth ieee international conference on data mining, pp. 413–422, 2008. W442W7302

[59] E. Fix and J. L. Hodges, "Discriminatory Analysis. Nonparametric Discrimination: Consistency Properties", International Statistical Review / Revue Internationale de Statistique, vol. 57, no. 3, pp. 238–247, 1989. W442W7302

[60] B. Ghojogh and M. Crowley, "Linear and quadratic discriminant analysis: Tutorial", arXiv preprint arXiv:1906.02590, 2019. W442W7302

9

Explainability and Interpretability Concepts for Edge AI Systems

Ovidiu Vermesan[1], Vincenzo Piuri[2], Fabio Scotti[2], Angelo Genovese[2], Ruggero Donida Labati[2], and Pasquale Coscia[2]

[1]SINTEF AS, Norway
[2]Università degli Studi di Milano, Italy

Abstract

The increased complexity of artificial intelligence (AI), machine learning (ML) and deep learning (DL) methods, models, and training data to satisfy industrial application needs has emphasised the need for AI model providing explainability and interpretability. Model Explainability aims to communicate the reasoning of AI/ML/DL technology to end users, while model interpretability focuses on in-powering model transparency so that users will understand precisely why and how a model generates its results.

Edge AI, which combines AI, Internet of Things (IoT) and edge computing to enable real-time collection, processing, analytics, and decision-making, introduces new challenges to acheiving explainable and interpretable methods. This is due to the compromises among performance, constrained resources, model complexity, power consumption, and the lack of benchmarking and standardisation in edge environments.

This chapter presents the state of play of AI explainability and interpretability methods and techniques, discussing different benchmarking approaches and highlighting the state-of-the-art development directions.

Keywords: edge AI, AI explainability, AI interpretability, explainable AI, XAI, trustworthy edge AI.

DOI: 10.1201/9788770041027-9   197

9.1 Introduction

Explainability and interpretability are terms used to describe how understandable edge artificial intelligence (AI), machine learning (ML), and deep learning (DL) models provide insight into their decision-making, as their complexity and opacity otherwise make it challenging to comprehend their behaviour. This is required to get confidence that edge AI models are dependable (e.g., reliable, resilient, secure, safe), trustworthy, and adhere to ethical principles appropriate to context, while ensuring that they are minimised. It is necessary to distinguish between explainability and interpretability to help developers and users in determining an AI/ML approach meets particular use cases.

Explainability is the ability to explain the decision-making process in terms that are understandable to the end user. An explainable model provides a clear and intuitive explanation of the decisions made, enabling users to understand why the model has produced a particular result; it focuses on why an algorithm has made a specific decision and how that decision can be justified. It requires a straightforward and intuitive presentation of information using an ontology familiar to the user. It is particularly valuable and beneficial in the case of deep neural networks, where the models are difficult to interpret due to the convoluted structure and complex internal interactions.

Interpretability is the ability to understand the decision-making process of an edge AI model. An interpretable edge AI model provides clear information about the relationship between inputs and outputs. An interpretable algorithm can be explained clearly and understandably by a person. Interpretability is essential to ensure that users will trust AI models.

While there are methods to explain the behaviour of models that are not inherently interpretable, interpretability serves as a gold standard for model explainability in a direct and transparent manner.

Superior AI explainability and interpretability come at the expense of performance, as illustrated Figure 9.1 [7]. When datasets are large, and the data are related to images or text, neural networks can meet the customer's AI/ML objective with high performance. For cases where complex methods are required to maximise performance, data scientists may focus on model explainability rather then of interpretability [7].

A conceptual workflow for the design of AI models which includes both interpretability and explainability is illustrated in Figure 9.2.

Interpretability is mostly associated with model training, evaluation, and quality assurance, while the explainability is a consideration of the deployed AI model.

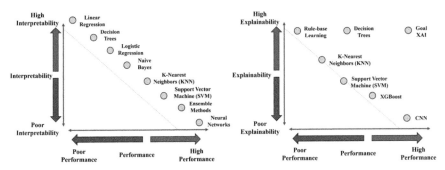

Figure 9.1 AI Interpretability and Explainability vs Performance for Common ML Algorithms (Adapted from [7])

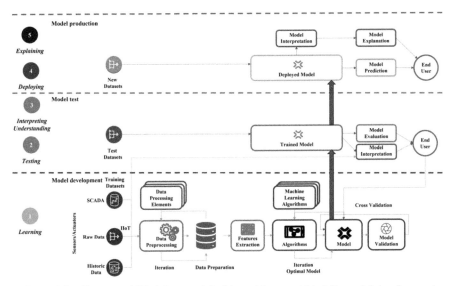

Figure 9.2 Conceptual Workflow explainable and interpretable ML model development

The European Union's Artificial Intelligence Act (AIA) [3] addresses AI explainability and interpretability. The AIA is a comprehensive regulatory framework that promotes transparency, accountability, and the protection of individual rights in the face of AI's growing influence, aiming to ensure the ethical and responsible use of AI. A significant proportion of current AI-based software falls within the scope of the AIA.

The European Parliament has amended the AIA by introducing Article 28 b, aligned with the 2019 OECD AI Principles [11], which states that AI "should be robust, secure, and safe throughout its lifecycle so that it functions appropriately and does not pose unreasonable safety risks." [12]. The new Article 28b features nine responsibilities for developers of foundation models. Of these nine obligations, the following three are the most relevant for AI designers;

Risk identification [Article 28b(2a)], which specifies that it is mandatory to identify and mitigate reasonably foreseeable risks (inaccuracy, discrimination, etc.) with the support of independent experts.

Testing and evaluation obliges AI providers to make adequate design choices to ensure that the foundation AI model achieves appropriate levels of performance, predictability, interpretability, corrigibility, safety, and cybersecurity. AI model functions are the building blocks for many downstream functions, so Article 28b(2c) aims to ensure that these meet the minimum standards and do not compromise systemic quality.

Documentation is an obligation for AI providers in the form of data sheets, model cards and intelligible use instructions. This is required to avoid that black box AI foundational models being deployed without knowing their processes or capabilities.

The documentation should include the following elements:

- A description of the data sources used in the development of the AI foundational model.
- An explanation of the capabilities and limitations of the foundational model, including reasonably foreseeable risks and the measures that have been taken to mitigate these, as well as the remaining unmitigated risks with an explanation of the motivation for which they could not be contained.
- A description of the training resources utilised by the foundation model, including the required computing power, the training time, and other relevant information related to the model's size, performance, and energy efficiency.
- A description of the model's performance based on public state-of-the-art industry benchmarking methods.
- A report and explanation of the results of relevant internal and external testing and optimisation of the model.

An overview of the responsibilities across the AI value chain according to the AIA is illustrated in Figure 9.3. The AIA provides a holistic approach to address the challenges posed by foundation models at different stages along the entire AI value chain. This approach considers that along the AI value

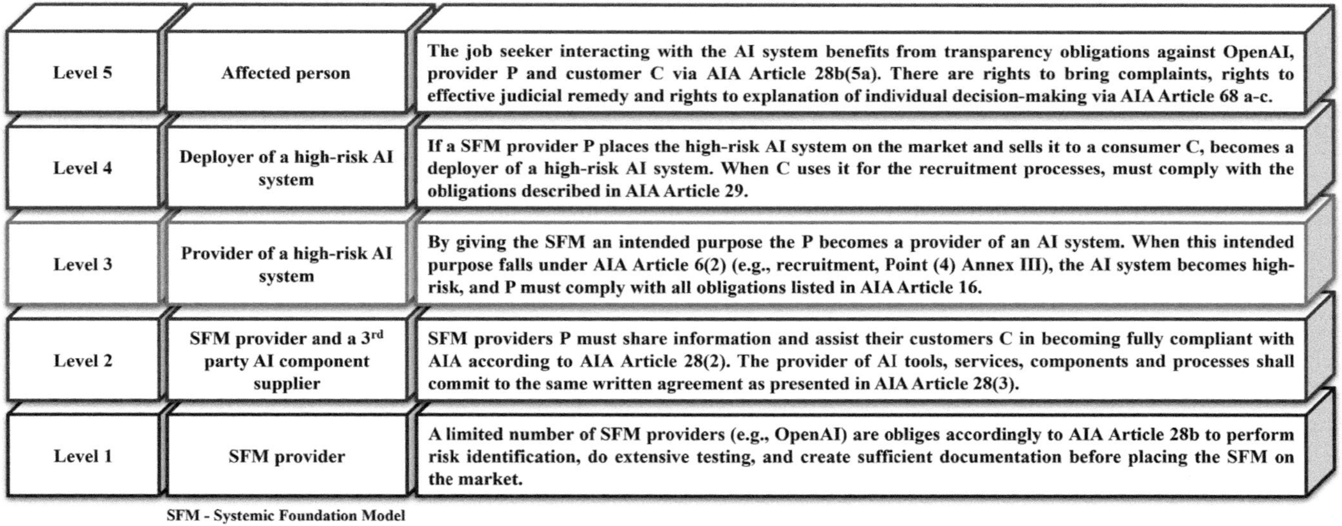

Level 5	Affected person	The job seeker interacting with the AI system benefits from transparency obligations against OpenAI, provider P and customer C via AIA Article 28b(5a). There are rights to bring complaints, rights to effective judicial remedy and rights to explanation of individual decision-making via AIA Article 68 a–c.
Level 4	Deployer of a high-risk AI system	If a SFM provider P places the high-risk AI system on the market and sells it to a consumer C, becomes a deployer of a high-risk AI system. When C uses it for the recruitment processes, must comply with the obligations described in AIA Article 29.
Level 3	Provider of a high-risk AI system	By giving the SFM an intended purpose the P becomes a provider of an AI system. When this intended purpose falls under AIA Article 6(2) (e.g., recruitment, Point (4) Annex III), the AI system becomes high-risk, and P must comply with all obligations listed in AIA Article 16.
Level 2	SFM provider and a 3rd party AI component supplier	SFM providers P must share information and assist their customers C in becoming fully compliant with AIA according to AIA Article 28(2). The provider of AI tools, services, components and processes shall commit to the same written agreement as presented in AIA Article 28(3).
Level 1	SFM provider	A limited number of SFM providers (e.g., OpenAI) are obliges accordingly to AIA Article 28b to perform risk identification, do extensive testing, and create sufficient documentation before placing the SFM on the market.

SFM - Systemic Foundation Model

Figure 9.3 Responsibilities Across the AI Value Chain

chain, multiple entities will supply tools, services and components, including data collection and pre-processing, model training, model retraining, model testing and evaluation, hardware/software integration. The complexity of the AI value chain requires transparency in a manner that permits traceability and explainability while making users aware that they are interacting with an AI system [3].

This chapter is organised as follows. Section 1 introduces the edge AI explainability and interpretability research area, including the proper definitions of the terms. Section 2 presents the goals of AI explainability and interpretability. Section 3 provides an overview of the state of the art of existing edge AI explainability and interpretability approaches, methods and techniques, and the actual advantages/disadvantages. Section 4 describes possible benchmarking techniques for edge AI explainability and interpretability to align with edge AI systems' trustworthiness requirements. Section 5 presents more detail on edge AI explainability and interpretability elements and specific issues. Section 6 describes the challenges, open issues, and future research directions for edge AI explainability and interpretability. Section 7 draws the conclusions.

9.2 AI Explainability and Interpretability Goals

Explainable and interpretable artificial intelligence enables trustworth predictive analytics, anomaly alerts, and decision-making. Data from edge devices can be analysed to predict maintenance for machines in industry and to optimise resource allocation in manufacturing. Effectively managing a distributed range of explainable systems to provide faithful computations on the data collected from edge devices is a fundamental challenge in deploying transparent edge-based AI applications. Creating effective solutions that can easily combine and accumulate decisions made by multiple models is still under development. It represents one of the key research areas to be investigated in the future [47]. Also aggregating explainability and interpretability in such composed systems represents a key challenge.

Over many years, researchers have primarily focused on enhancing model performance, relegating the intricate inner mechanisms that drive the output to a secondary analysis. Classical neural networks rely on millions of parameters (e.g., VGGNet has $\sim$138M parameters, and ResNet-152 has $\sim$60.3M parameters) [84]. Understanding the interconnections and communication pathways in these networks remains a challenging task. Furthermore, despite their remarkable performance, these models also exhibit vulnerabilities; object detectors and classification models, for example, can be easily deceived with slight alterations to input signals using adversarial examples [44], or decisions could be based on entirely incorrect features. Gender biases and stereotypes also pose challenges for Natural Language Processing (NLP) [45].

An understanding of the underlying mechanisms driving AI-driven model results has emerged as an imperative. This understanding is also a fundamental goal for human progress and for enhancing current AI-based systems. With the advent of new methodologies and large datasets, various sectors, including finance, transportation, healthcare, and security, have adopted approaches that are not only comprehensible but also endowed with an appropriate level of trustworthiness and effective oversight. For example, medical diagnosis systems usually employ visual explanations to provide support for their decisions, increasing the classification confidence [42]. The financial sector also heavily relies on interpretable methods for extracting trends and seasonalities from historical time series data [46].

In scenarios involving the proliferation of edge devices within a system, strategies that guarantee reliability, transparency, interoperability and foundational defence against vulnerabilities and errors become imperative, particularly in critical domains. The reliability of the analytics platform becomes crucial in these application scenarios. Autonomous systems equipped with the ability to perceive, learn, and make decisions represent the fundamental trajectory of future AI-based systems. Their actions must satisfy specific requirements and be explained in critical contexts.

Domains where interpretable systems find application span a diverse spectrum, for example:

Agriculture: Systems adept at extracting high-level insights from satellite images and remote sensors provide invaluable farming decision support. The possibility to expound upon the derived information is pivotal for informed decision-making [38].

Finance: Insurance companies and banks rely on automated systems to profile clients. These systems are pivotal in evaluating loan eligibility, demanding a transparent rationale for granting or withholding loans. Clear justifications are imperative for accountability and audit [36].

Industry and Autonomous Robots: Deploying automated systems to prevent human injuries requires the ability to proactively prevent individuals from specific actions. These systems must operate in a manner that absolves companies of liability for any unintended or improper action [37], while allowing post event analysis of any interventions that were performed.

Medical Diagnosis: Classifying magnetic resonance imaging (MRI) scans or histopathological images necessitates the elucidation of outcomes and the identification of causative factors. This is crucial for ensuring accurate diagnoses and comprehensible justifications for medical conclusions and interventions [35, 42].

Military and Security: Territorial defence and soldier training could considerably benefit from support systems that explain actions. These systems can enhance the efficiency of achieving goals, ensuring that tactical manoeuvres and training regimens are effective and comprehensively rationalised [39].

Recommendation Systems and Marketing: Typical applications consist of profiling users to support marketing endeavours that augments corporate revenues and facilitates the targeted promotion of products. Transparency in explaining these attributes fosters customer engagement and strategic decision-making [40].

Smart Cities: Aspects such as lighting, energy management, and traffic control within smart buildings and urban infrastructures are very applicable to AI. As the number of interconnected devices increases, AI-based frameworks must explain decisions regarding different aspects of human life (e.g., water supply, waste management, governance, etc.). Addressing cybersecurity and privacy challenges with explainable and interpretable methods is crucial for smart city development [43].

In addition, the General Data Protection Regulation (GDPR) [41], which codifies regulations on information privacy in the European Union and the European Economic Area, imposes legal obligations upon developers to elucidate decisions that hold the potential for impact on individuals. Finally, systems that inspire user confidence by being unambiguous and explainable are much more likely to be positively received and well engaged with.

9.3 AI Explainability and Interpretability Methods and Techniques

Highly accurate models are favoured over those that offer superior explainability but diminished accuracy, given that the primary objective of a machine learning system centres on its performance. However, it is not uncommon for these systems to be viewed as opaque by human evaluators, and the interpretation of their decision-making processes is often relegated to a subsidiary investigation.

Interpretability can enhance multiple aspects of a machine learning model. It can rectify biases learned during training, ensure that only meaningful variables contribute to the output, and measure robustness against adversarial perturbations. Sectors such as healthcare, finance, and security necessitate a profound understanding of ML models to uphold equity, responsibility, and transparency principles.

AI explainability and interpretability primarily focus on two aspects of an ML system: data and model. As illustrated in Figure 9.4, exploratory data analysis and visualisation represent important tools for gaining insights from data.

Dimensionality reduction techniques, such as PCA, ICA, t-SNE, LDA, and autoencoders, are used in cases involving many variables. These techniques convert high-dimensional data into a lower-dimensional form while preserving or extracting their internal structures.

Several frameworks implement data exploration and explanation techniques to express each feature's relevance through graphs, heatmaps, and various plots. Contrastive analyses provide interpretations that study the impact of features in achieving a desired output rather than solely focusing on the outcome itself.

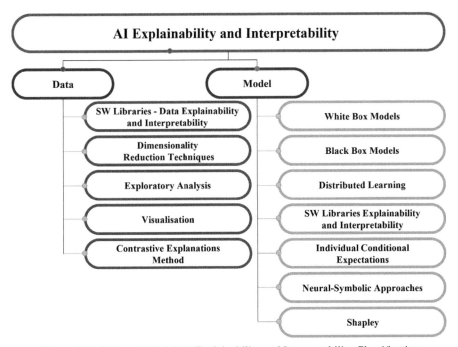

Figure 9.4   Data and Model AI Explainability and Interpretability Classification

While data explainability provides insights into the collected data, model explainability and interpretability focuses on the techniques used to understand the models. Specifically, explainable and interpretable models are categorised into transparent surrogate models, as illustrated in Figure 9.5.

Models classified as transparent inherently offer comprehensive insight through their intrinsic design or explicit processes aligned with the input data. Logistic or linear regression, decision trees, k-nearest neighbours and rule-based methods are examples of transparent models. This characteristic is mainly owned by ante-hoc methods.

Ante-hoc techniques allow embedding explainability into a model from the beginning. Post-hoc techniques enable models to be trained normally, with explainability only included at testing time.

Generalised additive models (GAMs) [54], for example, represent one of the first classes of nonparametric interpretable models, where the impact

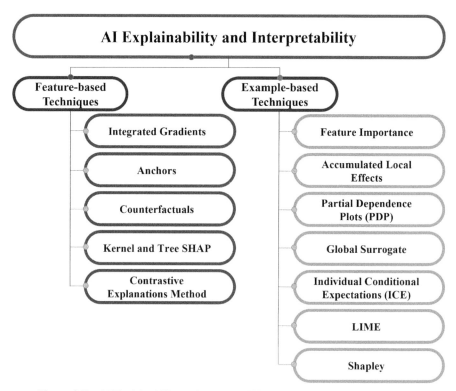

Figure 9.5   AI Explainability and Interpretability Model Approach Classification

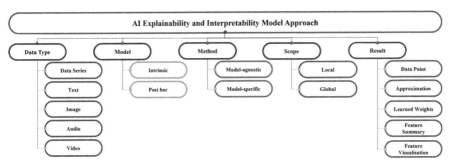

Figure 9.6 AI Explainability and Interpretability Model-Agnostic Approach Classification

of the examined variables is captured through smooth linear (or nonlinear) functions. Being additive, the effect, or impact of each variable can be measured independently from the others. Decision trees follow a tree-based logic, where control statements switch between specific paths to uncover rules behind decisions.

While computationally cheaper to evaluate, transparent models may not fulfil the performance criteria of the task at hand. Surrogate models use approximation criteria to emulate the operative dynamics of the primary model by assimilating the input-output relationship and exploiting fidelity measures [50] to evaluate their performance.

These models present fewer challenges in interpretation. They are created post-hoc and offer more flexibility and usability compared to the models they are built on top of. Post-hoc explainability refers to models that are not inherently interpretable by design and represent a class that encompasses diverse means to increase the explainability.

Post-hoc techniques offer valuable approximations of the inner workings or information flow to produce understandable representations using graphs, rule sets, score maps, or natural language.

While model-specific techniques extract explainable representations tailored to a particular learning algorithm or the internal structure of a model, model-agnostic techniques utilise model inputs and predictions to replicate the learning mechanism and generate explanations, as illustrated in Figure 9.6 and Figure 9.7.

Among model-specific techniques, feature importance highlights the impact of each feature on the decision.

Condition-based explanation defines oriented questions to allow the model to provide possible explanations with a set of conditions.

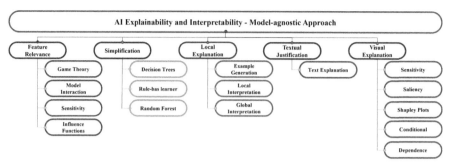

Figure 9.7   AI Explainability and Interpretability Model-Specific Approach Classification

Knowledge distillation methods [70] or rule-based learners [71, 72] also strongly rely on the original model.

Model-specific post-hoc explainable techniques cannot be employed with arbitrary models. In this circumstance, model-agnostic techniques can be considered since they involve conducting pairwise analyses of model inputs and predictions, aiming to comprehend the learning mechanism and generate explanations. This class, which does not make any assumptions about the model, includes visualisation-based techniques [73, 74], knowledge extraction [75, 76], and influence methods [77, 78]. Knowledge extraction provides a comprehensible representation of the model. Influence methods, instead, investigate the importance or resilience of hidden units by recording signal variations within the model.

The way explanations are presented is also inextricably linked to the nature of the data under examination. For instance, saliency, or attention, maps are prevalent to explain decisions derived from visual data (popular saliency methods are GradCAM [60], DeepLIFT [61] and SmoothGrad [62]); conversely, for textual data, specific segments of text that contribute to the resultant output are typically highlighted. Moreover, a predetermined set of rules can be applied to highlight the relevance of attributes in influencing the prediction.

Visual explanations represent one of the most important classes of methods used for classification, detection, and recognition tasks. Their success can be ascribed to the immediate representation of the decisions, highlighting what region of the input images generated that specific response. The medical domain, for example, extensively relies on these approaches [69].

These methods are typically used for visually understanding convolutional neural networks (CNNs) [66, 67, 68]. Most visual explanation techniques use backpropagation-based approaches that compute partial derivatives concerning each input feature or intermediate deep neural network layers [47] [48].

Another key distinction of the explanation generation processes relies on type of extracted explanations, which are representative of instances (local) or are broadly applicable (global). In this regard, local methods investigate the output of the models for specific samples and refer to a dynamic explanation process.

In this context, Local Interpretable Model-agnostic Explanations (LIME) [55] builds a surrogate model around the sample, which is easy to explain. A trade-off between unfaithfulness and the complexity of the model allows non-experts to interpret decisions by weighing the most critical parameters. Despite there being no guarantee that the surrogate models inherit the same properties as the original model, it is model-agnostic and only requires small perturbations to the input data.

Model Agnostic Supervised Local Explanations (MAPLE) [59] is a supervised neighbourhood approach that combines local linear models and ensembles of decision trees. SHAP (SHapley Additive exPlanations) [56] is another technique, based on game theory, used to explain the predicted output by computing the contribution of each input feature to the prediction.

Shapley values could refer to individual feature values or groups of feature values. For instance, pixels can be grouped into super pixels to explain an image. This method can be used both locally and globally. Other examples are counterfactual explanations [57].

Random Forest Feature Importance [63], Quasi Regression [64] and Global Sensitivity Analysis (GSA) [65] are examples of global methods that measure the importance of the features that contributed to the prediction highlighting their overall influence.

In this context, Partial Dependence Plots (PDPs) represent a class of visualisation-based techniques that define a global method able to visualise the effect of the values of a specific feature by marginalising all the other features.

Along with t-SNE, PCA and Quasi Regression, in these techniques the explanation is directly inferred from the black box model, compared to surrogate models. These methods are categorised as illustrated in Figure 9.8.

Whilst numerous methods were developed to explain the results, criteria to assess the explainability of a model are a fundamental and active area of research since several properties, such as casualty, target's belief, or trustiness, cannot be easily formalised [57].

Complexity and sparsity represent two critical aspects of evaluating a model to define its interpretability. The Predictive, Descriptive, Relevant (PDR) framework [58] proposes three desiderata for evaluating and

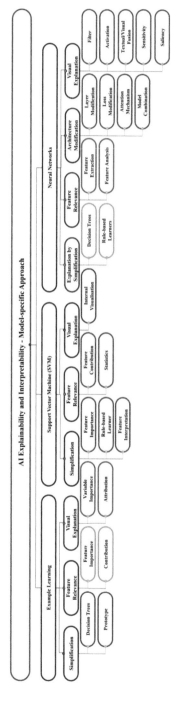

Figure 9.8   Feature- and Example-based AI Explainability and Interpretability Techniques

constructing interpretations: predictive accuracy, descriptive accuracy, and relevancy.

9.4 Benchmarking

The effectiveness of interpretable and explainable AI (XAI) techniques is influenced by various factors, including the user, usage context, model type, data characteristics, and desired form of explanation. Several approaches have been introduced in the literature to analyse and measure such effectiveness, the performance, and impact of interpretable and explainable AI techniques in real-life applications. However, the definition of a standard set of measures for evaluationg the effectiveness of interpretable and explainable AI techniques is still an open research problem, and there has yet to be an agreement on standard benchmarking methods.

The lack of accords stems from the fact that a qualitative human-based evaluation of the explanation is often necessary to assess the explanation quality. Nevertheless, several research trends are oriented towards the definition of quantitative approaches, enabling an automatic measurement of interpretable and explainable AI techniques, and allowing us to effectively compare different techniques [28].

It is therefore possible to distinguish two kinds of approaches to evaluate the effectiveness of interpretable and explainable AI techniques: *i)* quantitative evaluation methods, which involve creating an objective metric or benchmark to measure explanations without human involvement and that offer the advantage of facilitating comparisons between different explanation methods; *ii)* qualitative evaluation methods, which involve humans in evaluating explanations and permit evaluating the beneficial effects of interpretable and explainable AI methods from the users' perspective.

Quantitative evaluation approaches can be classified according to different taxonomies in the literature. As an example, [28] classifies evaluation approaches according to the type of application (images classifiers generating heatmaps, and natural language processing techniques). Moreover, recent studies propose the use of synthetically generated data with known properties to quantitatively evaluate the performance of interpretable and explainable AI methods [34]. However, generating realistic synthetic data with specific properties known a priori can be challenging for real application contexts. Together with classifying quantitative evaluation approaches, some work in the literature also review the measures used to evaluate their effectiveness. For example, [30] describes the following figures of assessments:

- *Fidelity* seeks to assess the accuracy of function f in emulating function b. Variations of fidelity exist, contingent upon the specific type of explainer being examined [31].

- *Stability* confirms that comparable instances yield consistent explanations. The assessment of stability can be accomplished using the Lipschitz constant [32].

- *Deletion* involves eliminating the features that were deemed important by the explanation method f, observing how the performance of b deteriorates as a result. One of the deletion methods is Faithfulness [32], which seeks to confirm whether the relevance scores truly reflect significance: higher importance values are anticipated for attributes that substantially influence the ultimate prediction.

- *Insertion* employs a complementary approach to deletion. Typically, both insertion and deletion evaluations are customised for specific types of explainers: Feature Importance explainers for tabular data, Saliency Maps for image data, and Sentence Highlighting for text data.

- *Monotonicity* [33] can be viewed as a manifestation of an insertion approach. It assesses the impact of b by systematically introducing each attribute in ascending order of importance. In this scenario, the anticipation is for the performance of the black-box model to progressively improve as more features are added, leading to monotonically increasing model performance.

- *Running time* is the computational time needed to provide interpretations or explanations. The running time of the technique used to explain the decisions made by the model in real time and cloud applications can be a critical factor. It is important for systems to provide interpretations or explanations in a timely manner.

Qualitative evaluation approaches can be classified according to whether they are designed to analyse explainable or interpretable AI methods. The qualitative analysis of explainable AI methods is mainly based on the statistical analysis of questionnaires submitted to human evaluation, which may be designed with different goals [29]:

- Evaluate the a priori goodness of explanations.
- Assess users' satisfaction with explanations.
- Uncover user's mental model of an AI system.
- Evaluate user's curiosity or need for explanations.
- Analyse the level of user's trust and reliance on the AI.
- Assess how the human-system work performs.

The qualitative analysis of interpretable AI methods is based on measures that can be systematised into three categories [30]:

- Functionally-grounded measures, which analyse the impact of the system in the considered application context.
- Application-grounded evaluation methods, which require evaluations performed by the set of human experts for which the system has been designed.
- Human-grounded measures, which assess interpretations using non-expert humans.

9.5 Edge AI Explainability and Interpretability

Integrating IoT, edge computing and AI can revolutionise how intelligent devices interact and enable a new era of innovative applications. By bringing computation, analytics, and connectivity closer to the data source, edge AI technologies reduce latency, enhance privacy, optimise bandwidth, and enable the online/offline operation.

Challenges such as limited computing resource, data quality and training, security and privacy, scalability, interoperability, ethical considerations, and explainability and interpretability must be addressed carefully. As these technology fields continue to advance, IoT, edge computing, and AI convergence are unlocking new opportunities, enabling intelligent decision-making and real-time insights at the edge.

AI at the edge extends ethical concerns about biased decision-making, algorithmic transparency, and accountability to that environment. As the number of intelligent edge devices increases, it is necessary to address ethical considerations and ensure that edge AI systems are fair, transparent, and accountable while edge AI models are explainable and interpretable. Compliance with legal regulations regarding data privacy, bias, and responsible AI usage is also crucial.

In the literature, there are only a limited number of studies on edge AI interpretability and explainability [80, 83]. Most of the work considers autonomous driving technologies [17], preventive healthcare applications [18, 80], and IoT [19].

Considering autonomous driving technologies, the study on edge AI interpretability and explainability regard different kinds of applications. There are methods for analysing images acquired from external cameras and Lidar sensors [20], and studies analysing the driver behaviour [21].

In preventive healthcare applications, interpretability and explainability techniques can detect possible health problems, as well as assist healthcare experts and family members in making critical healthcare decisions [22].

In the context of IoT devices, interpretability and explainability can be used to achieve heterogeneous goals according to the considered application scenario. For example, there are studies on edge AI interpretability and explainability for managing traffic [23], smart buildings [24], smart homes [25], environmental monitoring [26], and industrial control systems [27, 81, 82].

However, current studies on edge AI interpretability and explainability are limited to specific applications and do not propose a general approach for designing and developing interpretable and explainable AI technologies for the edge. This process is particularly challenging. In fact, developing edge AI solutions requires integrating edge AI hardware, software, AI stack building blocks techniques/methods/models and data addressed as a holistic edge AI design framework for the whole edge AI system.

Edge AI interpretability and explainability must apply to the edge AI model and data, as illustrated in Figure 9.4.

9.6 Challenges and Open Issues

Edge AI models are implemented and run on devices at the edge of a network, enabling real-time data processing and analysis. Edge processing is characterised by constrained computing, memory, power budget, and latency resources. Edge AI interpretability manages the extent to which a cause and effect can be observed within an edge AI system.

At the same time, explainability addresses how the internal mechanisms of an edge ML or DL system can be explained in human terms and representations. AI explainable and interpretable methods and techniques provide additional processing requirements and affect the overall performance of the AI-based systems implemented at the edge. This section presents several challenges, open issues, and future research directions that must be addressed for a successful edge AI deployment.

Edge AI model complexity vs interpretability and explainability is a challenge, considering AI decision-making must be transparent and understandable. Edge DL models are typically accurate but difficult to interpret. As a result, a trade-off between model complexity, interpretability, and explainability may be accepted. Complex models, such as edge deep neural networks (DNNs), capture convoluted patterns in data and provide prime performance. DNNs act as black boxes, making interpreting their behaviour or internal decisions challenging. AI models, such as decision trees or linear regression, are more straightforward and interpretable but offer lower performance on complex tasks and are more difficult to create.

The open issue is how to find the optimal balance to develop AI models that are powerful and robust enough to provide accurate results and yet sufficiently simple to be understandable. In many cases, this requires hybrid approaches, developing new edge AI interpretability and explainability techniques and methods, or accepting unavoidable trade offs in either explainability/interpretability or performance. In summary, achieving interpretability and explainability comes at the expense of edge AI model deployment. Simpler models that are easy to interpret may not perform as well as their complex replicas. Balancing the demand for explanation and interpretation with the requirement for models offering high-level performance is challenging.

Edge AI deployment and the management of AI models on many edge devices can be challenging considering the integration of edge AI explainable and interpretable methods, as it could be difficult to ensure that models perform optimally across all devices. Resource-constrained edge devices can also make running complex updates or retraining models challenging. This can be a significant problem as it is essential to monitor the performance of edge AI models and their explainable or interpretable surrogate models (twins) and implement regular maintenance, upgrades, and updates to prevent model degradation.

A lack of expertise in the field of edge AI explainability and interpretability will limit the adoption and deployment of edge AI. This can comprise the technical aspects of edge AI explainability and interpretability, such as how to build and optimise efficient explainable and interpretable models for edge devices and understanding the broader ramifications of using edge AI, such as real-time processing, latency, and security concerns. A lack of expertise can make it difficult to effectively design edge AI explainable and interpretable models and utilise them in edge AI applications to meet customers' requirements. It can also make it challenging for edge AI model providers and users to adequately evaluate the potential risks and benefits of using edge AI, limiting their ability to make informed decisions about possible adoption and deployment of edge AI.

Developing and deploying edge AI is a time-consuming and costly process and implies a trade-off between explainable and interpretable features and performance. Difficulties are associated with integrating edge AI explainable and interpretable models with edge devices, especially the ones with limited resources. The complexity and time associated with deploying edge AI explainable and interpretable models is a challenge, especially when dealing with large models, requiring extensive tuning and optimisation. Deploying, managing, and maintaining edge AI explainable and interpretable models on many edge devices is time-consuming and requires significant resources.

Updating and upgrading the edge AI explainable and interpretable models aligned with the improvements and advancements of edge AI models is essential to extend the lifetime of edge AI solutions. Adapting the features to the latest market advancements can be challenging, as edge AI solution providers must plan for incorporating the newest edge AI explainable and interpretable technology into their developments to stay competitive.

Edge AI explainability and interpretability is a relatively unexplored field with no standard definitions, established mature methods and techniques, best practices, or benchmarking methods. This can make it difficult for edge AI designers to know which approaches to adopt and how to measure their performance and efficiency. The choice of the approach depends on the specific edge AI model, its complexity, the intended solution, and the application's requirements. Combining different techniques may provide a more comprehensive interpretability and explainability solution for edge AI systems.

9.7 Conclusion

Explainable and interpretable AI models are applied to AI-based systems to complement them, facilitating the parallel use of data treatment, knowledge processing algorithms and analysable, and answerable implementations. This allows systems to simultaneously process relational and non-relational data from databases and sources that generate data in real-time, such as IoT sensors, and analyse the decision and outputs of the AI models.

The advancements in AI and edge AI require data analysis systems with AI algorithms and the parallel use of mathematical models for the creation of self-explanatory, self-answerable models that incorporate, for example, convolutional neural networks, deep symbolic learning, fuzzy logic, compartmental mathematical models, Bayesian networks, dynamic data assimilation models, and other models from the ML and DL domains.

The concepts of AI and edge AI explainability and interpretability are presented alongside emphasising that interpretability focuses on understanding the inner workings of the models. By contrast, explainability focuses on explaining the decisions made. As a result of the differences between the two concepts, interpretability requires more significant detailing than explainability.

The field of edge AI explainability and interpretability is evolving rapidly, and new approaches, methods and techniques are being developed to improve the explainability and interpretability of AI models and make them more transparent and more functional by improving visualisation methods, decomposition techniques, explanations based on examples, and ante-hoc and post-hoc approaches.

Edge AI involves deploying AI models on devices with inherent resource constraints, such as limited computing power, memory, and latency. Achieving a clear understanding of causality within these systems and making their internal workings and outputs comprehensible to humans often necessitates the use of hybrid approaches or the acceptance of trade-offs, with performance typically taking precedence.

The trade-offs are essential to edge AI explainability and interpretability as performance, energy consumption, complexity, and speed are constantly optimised against each other in resource-constrained edge devices. This is even more relevant considering the need for regular AI model updatability and upgradability.

Another essential consideration is that AI and edge AI models with advanced explainability or interpretability are mainly required in high-risk AI-based applications. Highly explainable/interpretable models can be used to assess AI-based systems by an independent third party and make another party accountable or liable while building trust between designers, developers, and users.

Currently, standardised definitions, mature methods, best practices, and benchmarking techniques are lacking in the field of edge AI explainability and interpretability. Nevertheless, there is an ongoing trend to explore comprehensive solutions that strike a balance between complexity, transparency, and the specific requirements of various applications. Addressing these challenges also requires the implementation of rigorous regulations and robust data quality validation. These efforts are becoming increasingly crucial as the networks of interconnected devices expand, adding complexity to the entire systems and emphasising the need for transparency.

This article attempts to classify and structure the existing concepts, offering the taxonomy needed to understand the multi-dimensionality of elements that must be considered, such as data (e.g., data type, data sets, and data use, encompassing – training, validation, testing, and inference, various AI model methods (e.g., model specific, model agnostic, etc.), extend (e.g., local, global) and the quality and behavioural properties (e.g., causality, transferability, fairness, informativeness, etc.).

In this context, edge AI explainability and interpretability solutions aim to ensure that AI models are transparent, accountable, and compliant with regulations, increasing user confidence and facilitating their adoption in various industries and applications.

Acknowledgements

This research was conducted as part of the EdgeAI "Edge AI Technologies for Optimised Performance Embedded Processing" project, which has received funding from KDT JU under grant agreement No 101097300. The KDT JU receives support from the European Union's Horizon Europe research and innovation program and Austria, Belgium, France, Greece, Italy, Latvia, Luxembourg, Netherlands, and Norway.

References

[1] A. Das and P. Rad. Opportunities and Challenges in Explainable Artificial Intelligence (XAI): A Survey. Available at: https://doi.org/10.485 50/arXiv.2006.11371

[2] F. K. Došilović, M. Brčić and N. Hlupić, "Explainable artificial intelligence: A survey," *2018 41st International Convention on Information and Communication Technology, Electronics and Microelectronics (MIPRO)*, Opatija, Croatia, 2018, pp. 0210-0215. Available at: https://doi.org/10.23919/MIPRO.2018.8400040

[3] European Parliament. Artificial Intelligence Act. P9_TA(2023)0236. Online at: https://www.europarl.europa.eu/doceo/document/TA-9-2 023-0236_EN.pdf

[4] K. Gade, S.C.Geyik, K. Kenthapadi, V. Mithal and A. Taly. Explainable AI in Industry, KDD '19: Proceedings of the 25th ACM SIGKDD International Conference on Knowledge Discovery & Data Mining, July 2019, pp. 3203–3204. Available at: https://doi.org/10.1145/3292500.33 32281

[5] D. Gunning, M. Stefik, J. Choi, T. Miller, S. Stumpf, and G-Z. Yang. "XAI—Explainable artificial intelligence." *Science robotics* 4, no. 37, 2019. Available at: https://openaccess.city.ac.uk/id/eprint/23405/8/

[6] S. R. Islam, W. Eberle, S. K. Ghafoor, and M. Ahmed. Explainable Artificial Intelligence Approaches: A Survey. Available at: https://doi.org/10.48550/arXiv.2101.09429

[7] J. King, B. Zhang, H. Mahboobi and S. Roy. "Model Explainability with AWS Artificial Intelligence and Machine Learning Solutions". AWS Whitepaper. 2021. Available at: https://docs.aws.amazon.com/pdfs /whitepapers/latest/model-explainability-aws-ai-ml/model-explainabil ity-aws-ai-ml.pdf?did=wp_card&trk=wp_card

[8] P. Linardatos, V. Papastefanopoulos, S. Kotsiantis. Explainable AI: A Review of Machine Learning Interpretability Methods. *Entropy*. 2021; 23(1):18. Available at: https://doi.org/10.3390/e23010018

[9] D. Minh, H.X. Wang, Y.F. Li, *et al*. Explainable artificial intelligence: a comprehensive review. *Artif Intell Rev* **55**, 3503–3568 (2022). Available at: https://doi.org/10.1007/s10462-021-10088-y

[10] M.Z. Naser, An engineer's guide to eXplainable Artificial Intelligence and Interpretable Machine Learning: Navigating causality, forced goodness, and the false perception of inference, Automation in Construction, Volume 129, 2021, 103821, ISSN 0926-5805. Available at: https://doi.org/10.1016/j.autcon.2021.103821

[11] OECD AI Principles overview. Available at: https://oecd.ai/en/ai-princi ples

[12] OECD AI Principle 1.4. Robustness, security and safety. Available at: https://oecd.ai/en/dashboards/ai-principles/P8

[13] G. P. Reddy and Y. V. P. Kumar, "Explainable AI (XAI): Explained," *2023 IEEE Open Conference of Electrical, Electronic and Information Sciences (eStream)*, Vilnius, Lithuania, 2023, pp. 1-6. Available at: https://doi.org/10.1109/eStream59056.2023.10134984

[14] D. Shin, The effects of explainability and causability on perception, trust, and acceptance: Implications for explainable AI, International Journal of Human-Computer Studies, Volume 146, 2021, 102551, ISSN 1071-5819. Available at: https://doi.org/10.1016/j.ijhcs.2020.102551

[15] V. Vishwarupe, P. M. Joshi, N. Mathias, S. Maheshwari, S. Mhaisalkar, and V. Pawar, Explainable AI and Interpretable Machine Learning: A Case Study in Perspective, Procedia Computer Science, Volume 204, 2022, pp. 869-876, ISSN 1877-0509. Available at: https://doi.org/10.1 016/j.procs.2022.08.105

[16] F. Xu, H. Uszkoreit, Y. Du, W, Fan, D. Zhao, and J. Zhu, J. (2019). Explainable AI: A Brief Survey on History, Research Areas, Approaches and Challenges. In: Tang, J., Kan, MY., Zhao, D., Li, S., Zan, H. (eds) Natural Language Processing and Chinese Computing. NLPCC 2019. Lecture Notes in Computer Science, vol 11839. Springer, Cham. Available at: https://doi.org/10.1007/978-3-030-32236-6_51

[17] D. Omeiza, H. Webb, M. Jirotka and L. Kunze, Explanations in Autonomous Driving: A Survey. *IEEE Transactions on Intelligent Transportation Systems*, vol. 23, no. 8, pp. 10142-10162, Aug. 2022, Available at: https://doi.org/10.1109/TITS.2021.3122865

[18] F. Di Martino, F. Delmastro. Explainable AI for clinical and remote health applications: a survey on tabular and time series data. *Artificial Intelligence Review*, vol. 56, pp. 5261–5315, 2023, Available at: https://doi.org/10.1007/s10462-022-10304-3

[19] İ. Kök, F. Y. Okay, Ö. Muyanlı and S. Özdemir, Explainable Artificial Intelligence (XAI) for Internet of Things: A Survey. *IEEE Internet of Things Journal*, vol. 10, no. 16, pp. 14764-14779, 15 Aug.15, 2023. Available at: https://doi.org/10.1109/JIOT.2023.3287678

[20] M. Abukmeil, A. Genovese, V. Piuri, F. Rundo and F. Scotti, "Towards Explainable Semantic Segmentation for Autonomous Driving Systems by Multi-Scale Variational Attention," *2021 IEEE International Conference on Autonomous Systems (ICAS)*, Montreal, QC, Canada, 2021, pp. 1-5, Available at: https://doi.org/10.1109/ICAS49788.2021.9551172

[21] M. P. S. Lorente, E. M. Lopez, L. A. Florez, A. L. Espino, J. A. I. Martínez, and A. S. de Miguel. Explaining Deep Learning-Based Driver Models, Applied *Sciences*, vol. 11, no. 8, p. 3321, Apr. 2021, Available at: https://doi.org/10.3390/app11083321

[22] R.-K. Sheu and M. S. Pardeshi, A Survey on Medical Explainable AI (XAI): Recent Progress, Explainability Approach, Human Interaction and Scoring System, *Sensors*, vol. 22, no. 20, p. 8068, Oct. 2022, Available at: https://doi.org/10.3390/s22208068

[23] A. R. Javed, W. Ahmed, S. Pandya, P. K. R. Maddikunta, M. Alazab, and T. R. Gadekallu, A Survey of Explainable Artificial Intelligence for Smart Cities, *Electronics*, vol. 12, no. 4, p. 1020, Feb. 2023, Available at: https://doi.org/10.3390/electronics12041020

[24] R. Machlev, L. Heistrene, M. Perl, K.Y. Levy, J. Belikov, S. Mannor, Y. Levron, Explainable Artificial Intelligence (XAI) techniques for energy and power systems: Review, challenges and opportunities, *Energy and AI*, vol. 9, 2022, Available at: https://doi.org/10.1016/j.egyai.2022.100169

[25] A. Dobrovolskis, E. Kazanavičius, and L. Kižauskienė, Building XAI-Based Agents for IoT Systems, Applied Sciences, vol. 13, no. 6, p. 4040, Mar. 2023, Available at https://doi.org/10.3390/app13064040

[26] I. Kalamaras, I. Xygonakis, K. Glykos, S. Akselsen, A. Munch-Ellingsen, H. T. Nguyen, A. J. Lepperod, K. Bach, K. Votis, D. Tzovaras. Visual analytics for exploring air quality data in an AI-enhanced IoT

environment. *Proceedings of the 11th International Conference on Management of Digital EcoSystems (MEDES '19).* Association for Computing Machinery, New York, NY, USA, 103–110, 2020, Available at https://doi.org/10.1145/3297662.3365816

[27] T.-T.-H. Le, A. T. Prihatno, Y. E. Oktian, H. Kang, and H. Kim, Exploring Local Explanation of Practical Industrial AI Applications: A Systematic Literature Review. *Applied Sciences*, vol. 13, no. 9, p. 5809, May 2023, Available at https://doi.org/10.3390/app13095809

[28] G. Ras, N. Xie, M. van Gerven, Marcel, D. Doran, Explainable Deep Learning: A Field Guide for the Uninitiated. *Journal of Artificial Intelligence Research*, vol. 73, pp.329-397, 2022, Available at https://doi.org/10.1613/jair.1.13200

[29] R. R. Hoffman, S. T. Mueller Shane, G. Klein, J. Litman, Measures for explainable AI: Explanation goodness, user satisfaction, mental models, curiosity, trust, and human-AI performance. *Frontiers in Computer Science*, vol. 5, 2023, Available at https://doi.org/10.3389/fcomp.2023.1096257

[30] F. Bodria, F. Giannotti, R. Guidotti, et al. Benchmarking and survey of explanation methods for black box models. *Data Mining and Knowledge Discovery*, vol.37, pp. 1719–1778, 2023, Available at https://doi.org/10.1007/s10618-023-00933-9

[31] R. Guidotti, A. Monreale, F. Giannotti, D. Pedreschi, S. Ruggieri and F. Turini, Factual and Counterfactual Explanations for Black Box Decision Making. *IEEE Intelligent Systems*, vol. 34, no. 6, pp. 14-23, 1 Nov.-Dec. 2019, Available at https://doi.org/10.1109/MIS.2019.2957223

[32] D. Alvarez-Melisì, T. S. Jaakkola. Towards robust interpretability with self-explaining neural networks. *Proc. of the 32nd International Conference on Neural Information Processing Systems (NIPS'18).* Curran Associates Inc., Red Hook, NY, USA, pp. 7786–7795, 2018, Available at https://dl.acm.org/doi/10.5555/3327757.3327875

[33] R. Luss, P.-Y. Chen, A. Dhurandhar, P. Sattigeri, Y. Zhang, K. Shanmugam, C.-C. Tu, Leveraging Latent Features for Local Explanations. *Proc. of the 27th ACM SIGKDD Conference on Knowledge Discovery & Data Mining (KDD '21).* Association for Computing Machinery, New York, NY, USA, pp. 1139–1149, 2021, Available at https://doi.org/10.1145/3447548.3467265

[34] R. Brandt, D. Raatjens, G. Gaydadjiev, Precise Benchmarking of Explainable AI Attribution Methods, *arXiv e-prints*, 2023, Available at https://doi.org/10.48550/arXiv.2308.03161

[35] A. Holzinger, G. Langs, H. Denk, K. Zatloukal, and H. Müller (2019). Causability and explainability of artificial intelligence in medicine. WIREs Data Mining and Knowledge Discovery, 9(4). Available at: https://doi.org/10.1002/widm.1312

[36] X.-Q. Chen, C.-Q. Ma, Y.-S. Ren, Y.-T. Lei, N.Q.A. Huynh, and S. Narayan (2023). Explainable artificial intelligence in finance: A bibliometric review. Finance Research Letters, 56, 104145. Available at: https://doi.org/10.1016/j.frl.2023.104145

[37] R. Setchi, M.B. Dehkordi, and J.S. Khan (2020). Explainable Robotics in Human-Robot Interactions. Procedia Computer Science, 176, 3057-3066. Available at: https://doi.org/10.1016/j.procs.2020.09.198

[38] M. Ryo (2022). Explainable artificial intelligence and interpretable machine learning for agricultural data analysis. Artificial Intelligence in Agriculture, 6, 257-265. Available at: https://doi.org/10.1016/j.aiia.2022.11.003

[39] D. Gunning, and D. Aha (2019). DARPA's Explainable Artificial Intelligence (XAI) Program. AI Magazine, 40(2), 44-58. Available at: https://doi.org/10.1609/aimag.v40i2.2850

[40] M. Liao, S.S. Sundar, and J.B. Walther (2022). User Trust in Recommendation Systems: A comparison of Content-Based, Collaborative and Demographic Filtering. In Proceedings of the 2022 CHI Conference on Human Factors in Computing Systems (CHI '22), Article 486, 1–14. Available at: https://doi.org/10.1145/3491102.3501936

[41] M. Ebers (2021). Regulating Explainable AI in the European Union. An Overview of the Current Legal Framework(s). Nordic Yearbook of Law and Informatics 2020: Law in the Era of Artificial Intelligence. Available at: http://dx.doi.org/10.2139/ssrn.3901732

[42] E. Tjoa, and C. Guan (2021). A Survey on Explainable Artificial Intelligence (XAI): Toward Medical XAI. IEEE Transactions on Neural Networks and Learning Systems, 32(11), 4793-4813. Available at: https://doi.org/10.1109/TNNLS.2020.3027314

[43] A. Kirimtat, O. Krejcar, A. Kertesz, and M.F. Tasgetiren (2020). Future Trends and Current State of Smart City Concepts: A Survey. IEEE Access, 8, 86448-86467. Available at: https://doi.org/10.1109/ACCESS.2020.2992441

[44] S. Thys, W. V. Ranst and T. Goedemé (2019). Fooling Automated Surveillance Cameras: Adversarial Patches to Attack Person Detection. IEEE/CVF Conference on Computer Vision and Pattern Recognition Workshops (CVPRW), Long Beach, CA, USA, 2019, pp. 49-55. Available at: https://doi.org/10.1109/CVPRW.2019.00012

[45] E. Balkir, S. Kiritchenko, I. Nejadgholi, and Kathleen Fraser (2022). Challenges in Applying Explainability Methods to Improve the Fairness of NLP Models. In Proceedings of the 2nd Workshop on Trustworthy Natural Language Processing (TrustNLP 2022), pages 80–92, Seattle, U.S.A.. Association for Computational Linguistics. Available at: http://dx.doi.org/10.18653/v1/2022.trustnlp-1.8

[46] Mandeep, A. Agarwal, A. Bhatia, A. Malhi, P. Kaler and H. S. Pannu. (2022). Machine Learning Based Explainable Financial Forecasting. 4th International Conference on Computer Communication and the Internet (ICCCI), Chiba, Japan, 2022, pp. 34-38. Available at: https://doi.org/10.1109/ICCCI55554.2022.9850272

[47] S. K. Jagatheesaperumal, Q. -V. Pham, R. Ruby, Z. Yang, C. Xu and Z. Zhang, "Explainable AI Over the Internet of Things (IoT): Overview, State-of-the-Art and Future Directions," in IEEE Open Journal of the Communications Society, vol. 3, pp. 2106-2136, 2022. Available at: https://doi.org/10.1109/OJCOMS.2022.3215676

[48] K. Simonyan, A. Vedaldi, and A. Zisserman (2014). Deep Inside Convolutional Networks: Visualising Image Classification Models and Saliency Maps. In 2nd International Conference on Learning Representations, ICLR 2014, Banff, AB, Canada, April 14-16, 2014, Workshop Track Proceedings. Available at: https://doi.org/10.48550/arXiv.1312.6034

[49] M.D. Zeiler, R. Fergus (2014). Visualizing and Understanding Convolutional Networks. In: Fleet, D., Pajdla, T., Schiele, B., Tuytelaars, T. (eds) Computer Vision – ECCV 2014. ECCV 2014. Lecture Notes in Computer Science, vol 8689. Springer, Cham. Available at: https://doi.org/10.1007/978-3-319-10590-1_53

[50] R. Guidotti, A. Monreale, S. Ruggieri, F. Turini, F. Giannotti, and D. Pedreschi (2018). A Survey of Methods for Explaining Black Box Models. ACM Comput. Surv. 51, 5, Article 93 (September 2019), 42 pages. Available at: https://doi.org/10.1145/3236009

[51] C. Molnar (2019). Interpretable machine learning. A Guide for Making Black Box Models Explainable. Available at: https://christophm.github.io/interpretable-ml-book/

[52] S. Ali, T. Abuhmed, S. El-Sappagh, K. Muhammad, J.M. Alonso-Moral, R. Confalonieri, R. Guidotti, J. Del Ser, N. Díaz-Rodríguez, F. Herrera (2023). Explainable Artificial Intelligence (XAI): What we know and what is left to attain Trustworthy Artificial Intelligence. Information Fusion, Volume 99, 2023, 101805, ISSN 1566-2535. Available at: https://doi.org/10.1016/j.inffus.2023.101805

[53] F. Doshi-Velez and B. Kim, "Towards A Rigorous Science of Interpretable Machine Learning", arXiv e-prints, 2017. Available at: https://doi.org/10.48550/arXiv.1702.08608

[54] T. Hastie and R. Tibshirani (1986). Generalized Additive Models. Statistical Science 1, no. 3, 297–310. Available at: http://www.jstor.org/stable/2245459

[55] M.T. Ribeiro, S. Singh, and C. Guestrin (2016). "Why should I trust you?: Explaining the Predictions of Any Classifier." In Proceedings of the 22nd ACM SIGKDD International Conference on Knowledge Discovery and Data Mining, Kdd San Francisco, CA, 1135–44. New York, NY: Association for Computing Machinery

[56] S. M. Lundberg and S.-I.Lee, "A Unified Approach to Interpreting Model Predictions", Advances in Neural Information Processing Systems (NIPS), 2017, pp. 4765–4774. Available at: https://dl.acm.org/doi/10.5555/3295222.3295230

[57] R. K. Mothilal, A. Sharma, and C. Tan, "Explaining machine learning classifiers through diverse counterfactual explanations", In Proceedings of the 2020 Conference on Fairness, Accountability, and Transparency (FAccT), 2020, pp. 607–617. Available at: https://doi.org/10.1145/3351095.3372850

[58] W.J. Murdoch, C. Singh, K. Kumbier, R. Abbasi-Asl, and B. Yu (2019). Definitions, methods, and applications in interpretable machine learning, Proceedings of the National Academy of Sciences, 2019. Available at: https://doi.org/10.1073/pnas.1900654116

[59] G. Plumb, D. Molitor, and A. Talwalkar (2018). Model Agnostic Supervised Local Explanations. Neural Information Processing Systems.

[60] R.R. Selvaraju, M. Cogswell, A. Das, R. Vedantam, D. Parikh, and D. Batra (2017). Grad-CAM: Visual Explanations from Deep Networks via Gradient-Based Localization. 2017 IEEE International Conference on Computer Vision (ICCV). Available at: https://doi.org/10.1109/ICCV.2017.74

[61] A. Shrikumar, P. Greenside, and A. Kundaje, "Learning important features through propagating activation differences", In Proceedings of the 34th International Conference on Machine Learning - Volume 70 (ICML), 2017, pp. 3145–3153. Available at: https://dl.acm.org/doi/10.5555/3305890.3306006

[62] D. Smilkov, N. Thorat, B. Kim, F. Viégas, and M. Wattenberg, "Smoothgrad: removing noise by adding noise", In Proceedings of the 2017 International Conference on Machine Learning (ICML), Workshop on

Visualization for Deep Learning. Available at: https://arxiv.org/abs/17 06.03825

[63] J. Friedman, T. Hastie, and R. Tibshirani (2001). The Elements Of Statistical Learning. Springer, New York. Available at: https://web.st anford.edu/Ÿhastie/ElemStatLearn/printings/ESLIIprint12.pdf

[64] T. Jiang and A.B. Owen, "Quasi-regression for visualization and interpretation of black box functions", 2002, Stanford University. Available at: https://artowen.su.domains/reports/qregvis.pdf

[65] P. Cortez and M.J. Embrechts (2011). Opening black box data mining models using sensitivity analysis. In Computational Intelligence and Data Mining (CIDM), 2011 IEEE Symposium on. IEEE. Available at: https://doi.org/10.1109/CIDM.2011.5949423

[66] Z. J. Wang et al., "CNN Explainer: Learning Convolutional Neural Networks with Interactive Visualization," in IEEE Transactions on Visualization and Computer Graphics, vol. 27, no. 2, pp. 1396-1406, Feb. 2021. Available at: https://doi.org/10.1109/TVCG.2020.3030418

[67] B. K. Iwana, R. Kuroki and S. Uchida, "Explaining Convolutional Neural Networks using Softmax Gradient Layer-wise Relevance Propagation," 2019 IEEE/CVF International Conference on Computer Vision Workshop (ICCVW), Seoul, Korea (South), 2019, pp. 4176-4185. Available at: https://doi.org/10.1109/ICCVW.2019.00513

[68] S. Albawi, T. A. Mohammed and S. Al-Zawi, "Understanding of a convolutional neural network", 2017 International Conference on Engineering and Technology (ICET), Antalya, Turkey, 2017, pp. 1–6. Available at: https://doi.org/10.1109/ICEngTechnol.2017.8308186

[69] T. Evans, C. O. Retzlaff, C. Geißler, M. Kargl, M. Plass, H. Müller, T.R. Kiehl, N. Zerbe, and A. Holzinger (2022). The explainability paradox: Challenges for xAI in digital pathology. Future Generation Computer Systems, Volume 133. Available at: https://doi.org/10.1016/j.future.202

[70] Hinton, G., Vinyals, O., and Dean, J.. Distilling the Knowledge in a Neural Network. ArXiv, 2015. Available at: 10.48550/arXiv.1503.02531

[71] Martens, D., Huysmans, J., Setiono, R., Vanthienen, J., Baesens, B. (2008). Rule Extraction from Support Vector Machines: An Overview of Issues and Application in Credit Scoring. In: Diederich, J. (eds) Rule Extraction from Support Vector Machines. Studies in Computational Intelligence, vol 80. Springer, Berlin, Heidelberg. Available at: https://doi.org/10.1007/978-3-540-75390-2_2

[72] Núñez, H., Angulo, C. & Català, A. Rule-Based Learning Systems for Support Vector Machines. Neural Process Lett 24, 1–18 (2006). Available at: https://doi.org/10.1007/s11063-006-9007-8

[73] P. Cortez and M. J. Embrechts, "Opening black box Data Mining models using Sensitivity Analysis," 2011 IEEE Symposium on Computational Intelligence and Data Mining (CIDM), Paris, France, 2011, pp. 341-348. Available at: https://doi.org/10.1109/CIDM.2011.5949423

[74] Alex Goldstein, Adam Kapelner, Justin Bleich & Emil Pitkin (2015) Peeking Inside the Black Box: Visualizing Statistical Learning With Plots of Individual Conditional Expectation, Journal of Computational and Graphical Statistics, 24:1, 44-65. Available at: https://doi.org/10.1080/10618600.2014.907095

[75] J. Tan, M. Ung, C. Cheng, and C.S. Greene, "Unsupervised feature construction and knowledge extraction from genome-wide assays of breast cancer with denoising autoencoders", Pacific Symposium on Biocomputing vol. 20, 2015, pp. 132–43. Available at: https://doi.org/10.1142/9789814644730_0014

[76] Goebel, R. et al. (2018). Explainable AI: The New 42? In: Holzinger, A., Kieseberg, P., Tjoa, A., Weippl, E. (eds) Machine Learning and Knowledge Extraction. CD-MAKE 2018. Lecture Notes in Computer Science (), vol 11015. Springer, Cham. Available at: https://doi.org/10.1007/978-3-319-99740-7_21

[77] A. Datta, S. Sen and Y. Zick, "Algorithmic Transparency via Quantitative Input Influence: Theory and Experiments with Learning Systems," 2016 IEEE Symposium on Security and Privacy (SP), San Jose, CA, USA, 2016, pp. 598-617, Available at: https://doi.org/10.1109/SP.2016.42

[78] P. W. Koh and P. Liang, "Understanding black-box predictions via influence functions", In Proceedings of the 34th International Conference on Machine Learning (ICML), Volume 70, pp. 1885–1894. Available at: https://dl.acm.org/doi/10.5555/3305381.3305576

[79] R. El Shawi, Y. Sherif, M. Al-Mallah and S. Sakr, "Interpretability in HealthCare A Comparative Study of Local Machine Learning Interpretability Techniques," 2019 IEEE 32nd International Symposium on Computer-Based Medical Systems (CBMS), Cordoba, Spain, 2019, pp. 275-280. Available at: https://doi.org/0.1109/CBMS.2019.00065.

[80] A. Yadu, P. K. Suhas and N. Sinha, "Class Specific Interpretability in CNN Using Causal Analysis," 2021 IEEE International Conference on Image Processing (ICIP), Anchorage, AK, USA, 2021, pp. 3702-3706. Available at: https://doi.org/10.1109/ICIP42928.2021.9506118.

[81] B. Malolan, A. Parekh and F. Kazi, "Explainable Deep-Fake Detection Using Visual Interpretability Methods," 2020 3rd International Conference on Information and Computer Technologies (ICICT), San Jose, CA, USA, 2020, pp. 289-293. Available at: https://doi.org/10.1109/ICICT50521.2020.00051.

[82] R. Jiang, Y. Xue and D. Zou, "Interpretability-Aware Industrial Anomaly Detection Using Autoencoders," in IEEE Access, vol. 11, pp. 60490-60500, 2023. Available at: https://doi.org/10.1109/ACCESS.2023.3286548.

[83] M. P. Neto and F. V. Paulovich, "Explainable Matrix - Visualization for Global and Local Interpretability of Random Forest Classification Ensembles," in IEEE Transactions on Visualization and Computer Graphics, vol. 27, no. 2, pp. 1427-1437, Feb. 2021. Available at: https://doi.org/10.1109/TVCG.2020.3030354.

[84] R. H. Valencia Tenorio, "Neuroevolved Binary Neural Networks". PhD thesis. The University of Auckland, 2020. Available at: https://researchspace.auckland.ac.nz/bitstream/handle/2292/57055/Valencia%20Tenorio-2020-thesis.pdf?sequence=1

Index

About the Editors

Dr. Ovidiu Vermesan holds a PhD degree in microelectronics and a Master of International Business (MIB) degree. He is Chief Scientist at SINTEF Digital, Oslo, Norway. His research interests are intelligent systems integration, mixed-signal embedded electronics, analogue neural networks, edge artificial intelligence and cognitive communication systems. Dr. Vermesan received SINTEF's 2003 award for research excellence for his work on implementing a biometric sensor system. He is currently working on projects addressing nanoelectronics, integrated sensor/actuator systems, communication, cyber-physical systems (CPSs) and the Industrial Internet of Things (IIoT), with applications in green mobility, energy, autonomous systems, and smart cities. He has authored or co-authored over 100 technical articles and conference papers. He is actively involved in the activities of the European partnership for Key Digital Technologies (KDT) Joint Undertaking (JU), now the Chips JU. He has coordinated and managed various national, EU and other international projects related to smart sensor systems, integrated electronics, electromobility and intelligent autonomous systems such as E3Car, POLLUX, CASTOR, IoE, MIRANDELA, IoF2020, AUTOPILOT, AutoDrive, ArchitectECA2030, AI4DI, AI4CSM. Dr. Vermesan actively participates in national, Horizon Europe and other international initiatives by coordinating and managing various projects. He is a member of the Alliance for IoT and Edge Computing Innovation (AIOTI) board. He is currently the coordinator of the Edge AI Technologies for Optimised Performance Embedded Processing (EdgeAI) project.

Dr. Dave Marples is Chief Scientist at Technolution in Gouda, NL where he is responsible for research and early-stage technical activities for the company. His primary research interests are in networked embedded systems in a Systems Engineering context. He regularly participates in EU Horizon and AENEAS research activities and is a member of the XECS Technical Experts Group. He is a member of the steering board of the IEEE Consumer Communications and Networking Conference (CCNC) and is a Fellow of the OSGi for services rendered as architecture chair and executive director.

He contributes to Open-Source projects and is the founder of orbcode.org. He was previously CS of the Internet Architecture Research laboratory at Telcordia (nee Bellcore) in Morristown, NJ where, amongst other things, he was responsible for the Networked Appliances research program. He was co-founder of the IEEE Communications Magazine Special Topic on Networked Appliances which runs today as the Internet of Things special topic. His PhD in Communications Systems is from Strathclyde University, Scotland and his BEng and MEng degrees in Electronics, Communication and Computer Engineering are from Bradford University, England. He is an Industrial Fellow of the Royal Society for the Exhibition of 1851 and lives in the UK with his partner, son and daughter. His hobbies include anything with an engine and Clay Pigeon missing.